Theory and Applications of Green Corrosion Inhibitors

Edited by

Inamuddin[1], Mohd Imran Ahamed[2],
Rajender Boddula[3] and Mohammad Luqman[4]

[1]Department of Applied Chemistry, Faculty of Engineering and Technology, Aligarh Muslim University, Aligarh-202 002, India

[2]Department of Chemistry, Faculty of Science, Aligarh Muslim University, Aligarh-202 002, India

[3]CAS Key Laboratory of Nanosystem and Hierarchical Fabrication, National Center for Nanoscience and Technology, Beijing 100190, PR China

[4]Chemical Engineering Department, College of Engineering, Taibah University, Yanbu Al-Bahr-83, Al-Bandar District, Postal Code: 41911, Kingdom of Saudi Arabia

Published by **Materials Research Forum LLC**
Millersville, PA 17551, USA

Published as part of the book series
Materials Research Foundations
Volume 86 (2021)
ISSN 2471-8890 (Print)
ISSN 2471-8904 (Online)

Print ISBN 978-1-64490-104-5
eBook ISBN 978-1-64490-105-2

Distributed worldwide by

Materials Research Forum LLC
105 Springdale Lane
Millersville, PA 17551
USA
https://www.mrforum.com

Manufactured in the United States of America
10 9 8 7 6 5 4 3 2 1

Table of Contents

Preface

Corrosion is a profoundly serious challenge to most of the industries around the globe because it can endanger lives, waste resources, and lead to economic loss. Therefore, corrosion of metallic materials is of significant interest to material scientists, chemists, and chemical engineers. Corrosion inhibitor is a chemical substance that meritoriously avoids or diminishes or controls the corrosion of exposed metal in a corrosive atmosphere. To protect metals and alloys from corrosion, several corrosion inhibitors are explored in the last decades. Due to the increasing environmental concerns and advancements in green chemistry principles, effective, financially viable and sustainable corrosion inhibitors need to be developed as smart coatings to control corrosion by avoiding toxic corrosion inhibitors. This book aims to provide a detailed analysis of research progress in sustainable corrosion inhibitors. It explores various kinds of green corrosion inhibitors and coatings via theoretical insights, characterization tools, and mechanisms. It also focuses on green corrosion inhibitors' applications, industrial and commercial technologies, current initiatives, critical challenges, and future directions. It is an essential guide for professionals, researchers, engineers, academics, and students in many fields including material science, chemistry, physics, polymer science, surface science and engineering.

Key features:

- Provides a comprehensive picture of green corrosion inhibitors
- Contains contributions from leading experts in the field of corrosion science
- Presents commercial applications of corrosion inhibitors

Chapter 1 presents the advances in the theoretical methods related to green corrosion inhibitors. The theoretical studies including quantum chemistry, molecular dynamics, simulation, and quantitative structure activity relationship (QSAR) on the green corrosion inhibitors for copper, aluminium and carbon steel are reviewed and discussed. Further research directions are also presented.

Chapter 2 examines the effects of natural resources and wastes on the corrosion resistance of metallic materials. In this study, the economic dimension of the process and future expectations are also examined.

Chapter 3 discusses factors attributed to the difference in the ability of metals to resist corrosion, and the different types of corrosion. The primary focus is on corrosion inhibitors extracted from green and biodegradable plants, also the use of biological wastes as corrosion inhibitors facilitating environmental protection.

Chapter 4 presents the state-of-the-art developments made with green corrosion inhibitors for preventing corrosion. It provides details of different electrochemical corrosion monitoring techniques including polarization technique, electrochemical impedance spectroscopy, electrochemical noise measurement and electrochemical quartz crystal microbalance.

Chapter 5 discusses several plant extracts (natural products) which have been explored as green corrosion inhibitors, as these are environmentally friendly, less expensive, widely available, reusable, and efficient. This chapter also provides a comprehensive study of technical applications of green corrosion inhibitors in various industries including reinforced concrete, coating, aircraft, oil and gas, acid pickling, and water industry.

Chapter 6 discusses the potential inhibiting nature of pyrazine and its derivatives towards corrosion of various metals and alloys in aggressive environments. Analytical tools such as gravimetric and electrochemical analysis exhibited good correlation with theoretical calculations to elucidate the pyrazine and its derivatives as green corrosion inhibitors.

Chapter 7 discusses different types of biological corrosion inhibitors which are effective for the prevention of reinforced concrete (RC) members. Mechanism of protection and efficiency of various microbial and botanical corrosion inhibiting extracts are elaborated in detail. Further, the pros and cons of both categories of inhibitors are discussed.

Chapter 8 presents details about the role of green corrosion inhibitors for protection of metals and alloys used in electronic devices. Advancements, and the protective role and adsorption mechanism of novel vapor-phase corrosion inhibitors (VPCI) as green corrosion inhibitors for electronic items, are discussed.

Editors

Inamuddin, Mohd Imran Ahamed, Rajender Boddula and Mohammad Luqman

Theory and Applications of Green Corrosion Inhibitors
Materials Research Foundations **86** (2021) 1-37

Materials Research Forum LLC
https://doi.org/10.21741/9781644901052-1

Chapter 1

Theoretical Insights in Green Corrosion Inhibitors

Guocai Tian

Faculty of Metallurgical and Energy Engineering, State Key Laboratory of Complex Nonferrous Metal Resources Clean Utilization, Kunming University of Science and Technolgy,Kunming, Yunnan Province, 650093, China

tiangc@iccas.ac.cn, tiangc01@163.com

Abstract

Metal corrosion is prevalent and has caused serious harm to the economy, industrial production, science and technology and other related fields. Adding corrosion inhibitors is an effective method to prevent metal corrosion. In recent years, considerable research progress has been achieved in green corrosion inhibitors for copper, carbon steel and aluminum. However, most works are mainly focused on the effect and efficiency of corrosion inhibitors, the microscopic mechanism of corrosion inhibitors is still unclear. Therefore, theoretical insights in green corrosion inhibitors are especially important. In the present chapter, we will present the advance and perspective on the theoretical study in green corrosion inhibitors. The theoretical studies, including quantitative structure activity relationship (QSAR), quantum chemistry, and molecular dynamics simulation on the green corrosion inhibitors for copper, carbon steel and aluminum are reviewed and discussed. Further research directions are also presented.

Keywords

Green Corrosion Inhibitors, Cooper, Aluminum, Carbon Steel, Quantum Chemistry, Molecular Dynamics Simulation, Absorption, Quantitative Structure-Activity Relationship (QSAR)

Contents

1. Introduction

Metal corrosion widely exists in the fields of industry, safety, economy, life and science and technology, and the harm and loss caused by it cannot be overlooked [1]. As reported in 2017, the global economic loss due to metal corrosion amounted to $ 2.5 trillion, accounting for about 3.4% of global gross domestic products [2]. In 2012, the cost of the corrosion of metal in the USA was approximately US $ 276 billion, which is equivalent to 16 times the economic loss caused by natural disasters (US $ 17 billion per year) [2,3]. The economic loss caused by metal corrosion is exceptionally large. In 2019, the direct economic losses caused by corrosion in USA, Japan and Canada will account for about 1% to 4% of the GDP of each country [3]. As far as China is concerned, losses due to corrosion in recent years and the cost of related protection of corrosion account for about 3% of GDP per year [4]. By using appropriate corrosion protection measures, corrosion costs can be cut by 25-40% [4,5]. Every year, large-scale equipment failure, building collapses and personal injuries are due to metal corrosion, especially for oil, natural gas and chemical pipeline equipment. The issue of metal corrosion has been a hot topic since the 1940s. Researchers have found that metal stress failure due to corrosion is a major factor in the destruction of large equipment [6].

Carbon steel, aluminum and copper are the three widely used metals in the industry, production and life. Carbon steel or mild steel is an alloy of iron and carbon, with good heat transfer, good electrical conductivity, soft texture and good magnetic properties. Aluminum is a silver white metal, light weight, with good ductility, conductivity, thermal conductivity, heat resistance and radiation resistance. The surface of aluminum in the air will form a dense oxide film, which makes aluminum have good corrosion resistance. Copper is a purplish red luster metal, slightly hard, extremely tough, wear-resistant, with good ductility, good thermal conductivity, conductivity and corrosion resistance. However, the corrosion of these three types of metal is not optimistic. The corrosion problem has caused huge economic losses. Therefore, the corrosion protection of carbon steel, aluminum and copper is of particular importance.

The corrosion of metals and its alloys in various fields in each country has caused serious problems in different fields and caused a series of major losses. Therefore, we must take effective corrosion protection measures. As latest reported from the National Association of Anticorrosive Engineers (NACE), the world's annual corrosion cost can be reduced by 15-35% (US $ 375-75 billion) by adopting appropriate anticorrosion technology [6].

To protection the corrosion of metals, many traditional anti-corrosion methods are used [6-8]. (1) Improving the nature of the metal. Adding alloy elements to the metals form an alloy to improve the corrosion resistance of metal, which can slow down or prevent the corrosion of the metal. (2) Forming a protective layer. Covering various protective layers on the metal surface is a highly effective and common method to separate the protected metal from the corrosive medium to prevent the corrosion of metal. Commonly used are metal phosphating treatment, metal oxidation and coating protection. (3) Improving the corrosive environment. Corrosion inhibitor is a common chemical substance, which can greatly reduce the rate of metal corrosion by adding it to the corrosive medium in a small amount. Owing to the small amount of corrosion inhibitor, simple and economical, it is a commonly used anti-corrosion method. (4) Electrochemical protection method. This method is an approach for preventing or mitigating metal corrosion by taking measures on metal equipment to make it become a cathode in corrosive batteries according to the principle of electrochemistry, including sacrificial anode and cathode protection.

Although these methods can suppress metal corrosion to a certain extent, most of the processes are relatively expensive, the processes are complex, and the slow-release efficiency is not high, which will have a definite impact on the environment. Therefore, it is urgent to find a green and effective method to replace these traditional processes. Adding a corrosion inhibitor is a green process that can effectively reduce corrosion.

Corrosion inhibitor refers to a mixture of chemical substances or a specified chemical compound, which can exist stably at an appropriate concentration in a corrosive environment, and can play a maximum role in slowing or preventing corrosion of metal materials [7,8]. Compared with other traditional processes, the addition of corrosion inhibitors has many advantages. It can be added directly in the system without the need for other additional equipment. According to different corrosion conditions, selecting a suitable corrosion inhibitor has excellent corrosion inhibition effects. There is no need to change the characteristics and structure of the protective metal. The amount of corrosion inhibitor is very small, and the use of a small concentration of corrosion inhibitor can play a good protective role, economically and effectively [5-8].

In recent years, the development and synthesis of new, efficient and environmentally friendly green corrosion inhibitor have attracted more and more attention. Some corrosion inhibitors represented by compounds such as nature products (e.g. plant extracts), ionic liquids (ILs), imidazolines, and amino acids have shown good environmental protection potential and currently received much attention. Compared with traditional corrosion inhibitors, green corrosion inhibitors are always renewable, inexpensive, ecologically acceptable, readily available and environmentally friendly and biodegradable. It does not include other toxic compounds or heavy metals. They are appropriate for different corrosion systems [4-12]. Various imidazoline, pyridine, ammonium, pyridazine, and benzimidazole type ionic liquids as effective corrosion inhibitors for carbon steel, copper, and aluminum were studied by surface analysis techniques, electrochemical tests, and electrochemical impedance spectroscopy [13-36]. In the present chapter we just focus on the performance of the ionic liquids as green corrosion inhibitor for carbon steel, copper, and aluminum. When ionic liquids were added to the systems, a protective film was formed on the metal surface through physical or chemical adsorption, which inhibits the reaction of the cathode or anode. Thereby, it is achieved a corrosion inhibition effect on the metal surface. The published work showed that for ionic liquids corrosion inhibitors formed with 1-butyl-3-methyl- imidazolium $[BMIM]^+$ and HSO_4^-, Cl^-, NO_3^-, BF_4^-, CH_3COO^- and $CF_3SO_3^-$, corrosion inhibition efficiency follow $HSO_4^- > Cl^-$ [15], $NO_3^- > BF_4^-$ [16], $CH_3COO^- > CF_3SO_3^- > Cl^-$ [33]. The inhibition efficiency of 1-methyl-3-methyl- imidazolium trifluoromethylsulfonate ([MIM]TfO) is better than that of 1-ethyl-3- methy- imidazolium tetrafluoroborate

([EMIM]BF$_4$ [23]). For ionic liquids corrosion inhibitors formed with 1-Hexyl-3-methylimidazolium ([HMIM]$^{+)}$ with TfO$^-$, I$^-$ and PF$_6^-$, the corrosion inhibition efficiency follows TfO$^-$ > I$^-$ > PF$_6^-$ [25]. On aluminum surface, pyridine, imidazole, and ester plasma liquids are effective corrosion inhibitors [37-42]. It was found that the longer the chain length of the cation is, the higher the corrosion inhibition of ionic liquid inhibitor formed by the same anion is [37-42]. These corrosion inhibitors can form a protective film on the aluminum surface. Long alkyl chains can slow down the corrosion process. For copper in acid solutions, various organic compounds such as benzotriazole, imidazole, phthalocyanine and their derivatives as corrosion inhibitors have been widely applied for copper and its alloys [43-47]. With the increase of the carbon chain, it was shown that the efficiency of corrosion inhibitor and the interaction between the benzotriazole and copper surface increase.

However, traditional method for choosing, development and design of corrosion inhibitors is a "trial and error" method. It needs a large number of experiments for screening, which has the disadvantages of high cost, long R & D cycle, and blindness. Again, the inhibitor mechanism is still unclear. Therefore, it is important to deeply reveal the inhibition mechanism of existing corrosion inhibitors to promote the research and development of economic, efficient and environmentally friendly green corrosion inhibitors, and to promote sustainable development of corrosion inhibitor for metal and its alloys.

Experimental research on the mechanism of corrosion inhibitor was carried out earlier, and many valuable results were obtained. Various experimental research methods and detection technologies provide reliable means for design, development and synthesis new type inhibitors for the corrosion protection of metal. The experimental research methods of the corrosion inhibition mechanism mainly include surface analysis techniques, electrochemical tests, and electrochemical impedance spectroscopy. Weight loss method and Tafel curve extrapolation method are the two most commonly used methods to study the corrosion inhibition characteristics. However, due to the limitations of the technology itself, numerous experimental techniques and methods are not yet perfected. For example, the weight loss method cannot reflect the morphology of corrosion. Although the polarization resistance method can measure the instantaneous rate of metal corrosion rapidly, it can't judge the degree of corrosion inhibitor to anode and anode process. Regardless of the fact that the photo-electrochemical method can study the electrochemical reaction process from a microscopic perspective, it has the disadvantage of insufficient reflection of local electronic properties and corrosion processes on the electrode surface. The AC impedance method has unique advantages in assessing the influence of inhibitors on the metal corrosion process, but the slow-developing spectral analysis technology has severely restricted the application and development of this technology.

With the gradual maturity of technology in recent years, many surface analysis methods such as scanning tunneling microscope (STM) [48], auger electron spectroscopy (AES) [49], surface enhanced Raman scattering (SERS) [50], and X-ray photoelectron spectroscopy (XPS) [51] are increasingly applied to study the mechanisms of corrosion inhibitor. These methods can be used to conveniently and accurately measure the elemental composition of the inhibitor film, analyze the molecular structure of organic compounds adsorbed on metal or material surface, and explore the behavior of adsorption and desorption of the inhibitor. However, there are also limitations and deficiencies in various surface analysis technologies, and there are still many problems to be resolved in the research of corrosion and corrosion inhibitor. For example, the STM method are used to study the stability, structure and influencing factors of the film which formed by the inhibitor molecules on inert metal or less active metal such as Au, Pt, Ag, and Cu surface. However, for the more active metal such as Fe, Al, and Zn, because it is easily oxidized in air or solution, it is difficult to obtain a flat and clean surface. Therefore, the adsorption layer on the surface of these active metals can only be conducted in ultra-high vacuum. The AES method needs to be performed under vacuum conditions, so it is not feasible to obtain a surface state corresponding to the actual corrosion condition. The SERS method can directly give information about the molecular structure of the adsorption, but the experiment can only be completed on the surface of metal (Au, Cu, Li, Na, etc.) with high SERS effect. At the same time, in order to avoid interference with the SESR spectrum, non-paramagnetic substances must be selected as the adsorbent, which greatly limits the application of the SESR technology. Therefore, the application of surface analysis technology in corrosion inhibitor needs further study.

Due to the limitations of experimental methods and surface analysis techniques itself, some detailed information about the corrosion inhibition process cannot be revealed, such as the quantitative description of the inhibitor molecule-metal atom binding, the determination of the inhibitor group, and the acquisition of information on the process of the inhibitor film formation by adsorption, as well as microscopic diffusion behaviors of the corrosion particles in the corrosion inhibitor film, and so on. In addition, the correctness of the corrosion inhibition process and mechanism based on qualitative prediction based on experimental and basic knowledge in the past also needs to be further confirmed. In order to better understand the nature and characteristics of the corrosion inhibition process, we should develop or apply other effective methods to solve these problems. With the rapid development of computer technology, molecular simulation technology has been a common approach to study the complex problems at micro-scale in recent years. It can provide detailed supplementary experiments on the molecular structure and adsorption process of inhibitor information, which creates conditions for a deeper discussion of the corrosion inhibition mechanism. It is the future development trend of the corrosion inhibitor technology field.

2. Theoretical methods used in green corrosion inhibitors

Molecular simulation methods have been widely and successfully applied to reveal inhibition mechanism and behavior in terms of structure-activity relationship, adsorption law, mechanism of film formation and corrosion inhibition of the corrosion inhibitors. Many valuable research results have been obtained, which provides strong technical support for in-depth understanding of corrosion inhibition mechanism and progress of inhibitor molecular design. Theoretical research on green corrosion inhibitors mainly include quantum chemistry, quantitative structure-activity relationship and molecular simulation, as well their combination.

2.1   Quantum chemistry methods

The quantum chemistry method mainly studies the electronic structure of atoms, molecules, and crystals, chemical bonds, and their characteristics of various spectra, spectroscopy, and electron spectroscopy by solving Schrödinger equation of the system. Due to different methods of solving the Schrödinger equation, various calculation methods have been generated. Quantum chemical calculation methods commonly used in early corrosion research include CNDO, CNDO/2, MINDO, and MINDO/3. These methods are all semi-empirical molecular orbital methods. Although the reliability of the calculation results is not ideal, they are continuously improved and gradually improved in the long-term application process. At present, they still play an important part in the research of corrosion inhibition mechanism. From 1980s to 1990s, a new type of quantum chemical method, Density Functional Theory (DFT) [52-54] developed rapidly. It switched the characteristics of the molecular orbital method based on the wave function and switched to electron density function. Compared with the molecular orbital method, the density functional theory has the characteristics of minute calculation scaling and high accuracy. Nowadays, it is more and more commonly applied to reveal the inhibition behavior and mechanism and to design the inhibitor. Research on quantum chemistry of corrosion inhibition mechanism in the United States, Britain, Mexico and other countries has been carried out earlier, and has achieved many valuable results, and gradually expanded to the direction of molecular design.

In 1971, Vosta and Eliasek [55] used quantum chemistry to study corrosion inhibition mechanism and pioneered the quantum corrosion electrochemistry. Since then, many scholars have used quantum chemistry to explore corrosion inhibition mechanism. The hot issue is mainly to study the correlations between the structure information of organic corrosion inhibitors and the efficiency of corrosion inhibitor. Internal characteristic parameters are obtained by quantum chemistry calculation, such as charge distribution, LUMO (lowest unoccupied molecular orbits) energy, dipole distance, HOMO (highest occupied molecular orbits) energy, and Fukui index etc. These parameters are applied to analyze the performance of corrosion inhibition, and to explore the possible inhibition mechanism. They are also correlated with inhibition efficiency (experimental data) of corrosion inhibitor.

Quantum chemistry methods are often used to perform geometry optimization and vibrational frequency analysis to obtain the stable geometry of the target systems. Various molecular structure parameters from quantum chemical calculation including hardness (η), LUMO energy (E_{LUMO}), softness (σ), dipole moment (μ), HOMO energy (E_{HOMO}), total energy, the energy gap $\Delta E = E_{LUMO} - E_{HOMO}$, polarizability ($P$), ionization potential (I), electronegativity (χ), electrons transferred number (ΔN), electron affinity (A) and electrophilicity (ω) can be obtained from quantum calculation. These parameters are used to reveal the reaction activity of the target systems.

Based on Koopmans theorem [56], in the vacuum conditions, the ionization potential (I) is described by the negative value of LUMO energy (E_{LUMO}),

$$I = -E_{HOMO},$$
(1)

whereas the electron affinity (A) is approximately defined as

$$A = -E_{LUMO}.$$
(2)

According to the conceptual density functional theory [57], the hardness (η) and electronegativity (χ) at a constant external potential $v(r)$ can be described by electronic energy (E) and electron number (N) [58-59] by

$$\chi = -\left(\frac{\partial E}{\partial N}\right)_{V(r)}$$
(3)

$$\eta = \frac{1}{2}\left(\frac{\partial^2 E}{\partial N^2}\right)_{V(r)}$$
(4)

Pearson and Parr [60] proved that Eq. (3) and (4) can be calculated from electron affinity A and ionization energy I of compounds, namely,

$$\chi = \frac{I + A}{2} \tag{5}$$

$$\eta = \frac{I - A}{2} \ . \tag{6}$$

According to Koopmans theorem [56], the electronegativity (χ) and hardness (η) can be calculated with E_{HOMO} and E_{LUMO} by Eq. (7) and (8),

$$\chi = -\frac{E_{LUMO} + E_{HOMO}}{2} \tag{7}$$

$$\eta = \frac{E_{LUMO} - E_{HOMO}}{2} \ . \tag{8}$$

The softness (σ) is proposed by R.G. Pearson [61] and can be described by the multiplicative inverse of hardness

$$\sigma = \frac{1}{\eta} \ . \tag{9}$$

Electrophilicity (ω) was introduced by parr [61], and was regarded as a way of determining molecular reactivity toward attracting electrons from a chemical specie, which is given by

$$\omega = \frac{\chi^2}{2\eta} \ . \tag{10}$$

The electron transferred (ΔN) that an inhibitor molecule transfers to metal surface was given by [62]

$$\Delta N = \frac{\chi_{Fe} - \chi_{inh}}{2\left(\eta_{Fe} + \eta_{ihh}\right)} \ , \tag{11}$$

χ_{inh} and χ_{Fe} are the absolute value of electronegativity of inhibitor and metal iron, η_{inh} and η_{Fe} are the absolute value of hardness for a inhibitors and metal iron.

The polarizability is a physical quantity to measure the response ability of molecule to electric field and the ability to obtain the electric-dipole moment. The observable mean value of polarizability [63] is given as follow

$$P = \frac{1}{3}\left(\alpha_{xx} + \alpha_{yy} + \alpha_{zz}\right),\tag{12}$$

where α_{xx}, α_{yy} and α_{zz} represent the polarizability in x, y and z direction respectively.

The local reactivity of chemical compounds is described by Fukui functions f_k [64]. In a constant external potential $v(r)$, the f_k was defined by electrons number N and the electronic density $\rho(r)$ [65].

$$f_k = \left(\frac{\partial \rho(r)}{\partial N}\right)_{v(r)}\tag{13}$$

The Fukui function can be given approximately as

$$f_k^+ = q_k(N+1) - q_k(N),\tag{14}$$

$$f_k^- = q_k(N) - q_k(N-1),\tag{15}$$

where $q_k(N+1)$, $q_k(N-1)$ and $q_k(N)$ are gross charge of atom k of cationic species, anionic, and neutral respectively. N is the number of electron in neutral species.

HOMO electron density distribution commonly determines the possible sites or regions in molecules associated with donating electrons to other molecules or ions, while the LUMO electron density distribution determines the possible sites or zones in molecules associated with accepting electrons from other molecules or ions. The energy gap is a key physical quantity to describe the relative activity of molecules. In general, a low ΔE means that the adsorption of a molecule on the metal surface is much easier, and corresponding corrosion inhibition efficiency is higher [66]. Molecular polarizability is an important quantity to assess corrosion inhibition performance. In general, the higher value of polarizability means that the molecule can be adsorbed easily on the metal surface [63]. Hardness (η) and global softness (σ) can be related to the selectivity and reactivity of the molecule, and the adsorption often occurs at the part of the molecule with lower η and higher σ values [67]. Electronegativity (χ) and electrophilicity (ω) are widely used in the prediction of inhibition performance of molecules and chemical behavior of molecules on the surface of metal. Inhibitor molecule with low electronegativity and electrophilicity values is a good corrosion inhibitor [68]. Fukui function $f(r)$ is usually applied to confirm the active sites of the compounds and predict the local relative reactivity in the compounds [68]. In general, electrophilic attack is more like to take place at the highest f^- value site in a compound, while nucleophilic attacks sites represent a highest value of f^+ [69].

Gómez et al. [70] calculated the electronegativity, hardening degree, softening degree and Fukui index of the four amino compounds using density functional theory, and evaluated the corrosion inhibition performance of the four amino compounds based on these parameter values. Vosta [55] reported that the higher the electron density of nitrogen atom in aniline derivative is, the lower the corrosion rate and the better the linear relationship is.

2.2   Quantitative structure-activity relationships

Quantitative structure activity relationship (QSAR) is the most commonly used method of drug design [71,72]. Quantitative structure-activity method is to determine the quantitative relationship between the physiological activity and physical and chemical properties of a series of compounds through some mathematical and statistical methods. It can predict the physiological activity or certain properties of the compound, and guide us to develop compounds with higher activity. As early as 1867, Crum-Brow and Fraser proposed the concept of structure-activity relationship [71,72]. Around 1900, Overton and Meyer et al. proposed a lipid theory of anesthesia, that is, the activity of anesthetics with different chemical structures increases as the lipid-water partition coefficient increases. This may be the earliest proposed model of quantitative distribution relationship between physiological activity and physicochemical properties of compounds. But only in recent decades, especially after the Hansch method was proposed, with the introduction of multivariate analytical techniques and development of computer technology, quantitative structure-activity relationship methods have since become an indispensable tool in design and development of the drug, compound or material.

With the development of computer software and hardware technology, in recent years, QSAR method has become an indispensable tool in drug design and drug development, and has been successfully and widely used in many fields such as pharmacy, materials science, environmental chemistry, bioinformatics and so on. The most commonly used QSAR method for corrosion inhibitors, is the equation proposed by Lukocits [73] for establishing the correlation between molecular structure parameters and the experimental inhibition efficiency of corrosion inhibitors with different concentrations. Based on the

Langmuir adsorption isotherm equation, the linear equations (in Eq. (16)) and nonlinear equations (in Eq.(17)) of structure-activity relationship between the structure parameters of inhibitor molecules and their inhibition efficiency were derived, namely,

$$IE = Ax_j C_i + B \qquad (16)$$

$$IE = \frac{\left(Ax_j + B\right)C_i}{1 + \left(Ax_j + B\right)C_i} \quad . \qquad (17)$$

Where A and B are the fitting coefficients, *IE* is corrosion inhibition efficiency, x_j are structure parameters of inhibitors obtained from quantum chemical calculation, and C_i is inhibitor concentration.

Abdul-Ahad and Al-Madfat [74] used nine para-substituted aniline compounds as research objects, calculated their physical and chemical parameters using the CNDO/2 method, and established a mathematical model between these parameters and the corrosion rate. Bentiss et al. [75] and others combined the experimental method with theoretical calculations, and conducted a structure-effect analysis of the molecular front-line orbital energy and dipole moment to establish a linear resistance model to study the retardation of triazole and oxazole in 1M HCl. Zhang et al. [76] and others introduced the molecular topological index into the QSAR to establish the relationship between inhibition efficiency and the molecular structure parameters of a series of N-containing corrosion inhibitors. They used a combination of structure parameters from quantum chemical calculation and molecular topological index to establish the correlation relationship. El Sayed H. El Ashry [77,78] and others used quantum chemical methods to calculate the front-line orbital energy, dipole distance, and the total negative charge of 19 azole molecules, and established these physical and chemical parameters and corrosion inhibition efficiency. The mathematical relationship model between them; then semi-empirical algorithms were used to calculate the front-track orbital energy, dipole distance, ester-water partition coefficient, the total negative charge, intrinsic molecular volume, and other parameters of twenty benzimidazole molecules, and statistical analysis approaches were used. A linear correlation equation of corrosion inhibition performance with their molecular structure parameters. Hu et al. [79] utilized 15 kinds of imidazoline corrosion inhibitors as research objects, established a QSAR model between the molecular structure of corrosion inhibitors and inhibition efficiency in H_2S and CO_2 coexisting corrosion environments. The scientifically, stability and prediction ability of the model has been systematically researched. Based on this, four additional corrosion inhibitor molecules with good resistance to H_2S and CO_2 corrosion have been designed and synthesized.

2.3   Molecular dynamics simulation

Molecular dynamics (MD) and Monte Carlo (MC) method [11] are commonly applied approaches to study the microstructure, interaction of particles, thermodynamic properties, and dynamic properties in complex system. MD simulation is a molecular simulation method based on classical mechanics, which are equivalent to calculating the translation and equilibrium transfer properties of a particle. Using statistical principles, the average value of particle momentum versus time can be achieved. Using Newton's equation, particles can be obtained velocity and position at different times, and periodic boundary conditions are added to simulate the actual system. Molecular dynamics simulation can be utilized to analyze the interaction energy between molecules and study the microstructure, thermodynamic and dynamic properties of matter.

The basic principle of MC simulation is the principle of random molecular calculation, which does not contain molecular momentum information. Through these two simulation methods, the adsorption energy of ionic liquid inhibitors on metals and alloys can be obtained, the size of the inhibitor's corrosion inhibition performance, and the adsorption functional position can be obtained [34-37].

In the simulation of corrosion inhibitors, the interaction of corrosion inhibitors with the metal surface are described by binding energy $E_{binding}$ or adsorption energy $E_{adsorption}$. In vacuum binding energy $E_{binding}$ or adsorption energy $E_{adsorption}$ is given by [80]

$$E_{adsorption} = E_{total} - \left(E_{surface} + E_{inhibitor} \right)$$

$$E_{binding} = - E_{adsorption},$$

(18)

where $E_{total}, E_{surface}$ and $E_{inhibitor}$ are the energy of metal surface-inhibitor complex, pure metal surface and inhibitor molecule obtained from MD simulations respectively. In solution medium, $E_{binding}$ or $E_{adsorption}$ can be calculated by [81]

$$E_{adsorption} = E_{total} - \left(E_{surface+solution} + E_{inhibitor+solution} \right) + E_{solution}$$

$$E_{binding} = - E_{adsorption},$$

(19)

where $E_{solution}$ $E_{surface+solution}$, and $E_{inhibitor+solution}$ are the energy of the solution, metal surface-solution, inhibitor molecule-solution system obtained from MD simulations respectively.

The role of the corrosion inhibitor and metal interface is very complicated. In the quantum chemistry study of the corrosion inhibition mechanism, it is not enough just to consider only the isolated inhibitor molecules. It necessary to comprehensively considers the situation of the metal interface. So researchers turned their attention to the research of inhibitor-metal interface system. Ramachandran et al. [82] analyzed the bonding between imidazoline oleate and iron clusters. The calculation results show that imidazoline oleate as a Lewis base has strong bonding with Fe^{3+}. It was established that the adsorption of imidazoline oleate on the surface of iron clusters was stronger than that of water. Luo Daoming [83] calculated the binding energy and bond length of adsorption for the amine/iron system by the CNDO/2 method, and pointed out that the larger adsorption binding energy (absolute value), the stronger the interaction between the iron interface and the amine molecule, and the higher the corrosion inhibition efficiency. Lawrence [84] constructed a cluster model of 28 iron/aluminum atoms to approximate a metal surface. The DFT was utilized to study the interaction between a series of aniline terpolymers on iron/aluminum surface. The binding energy of metal surface with molecule, the orbital energy, electron distribution, ionization energy and other parameters of the molecule were applied to reveal the behavior of adsorption of a molecule on the surface of metal. The corrosion inhibition properties of various aniline terpolymers were analyzed and evaluated. These studies revealed the reactivity and selectivity of the inhibitor molecules and their adsorption behavior and regularity on the surface of metal, which is very importance for understanding the mechanism of corrosion inhibitor. However, the work of using corrosion inhibitor-metal interface system to study the inhibition mechanism is relatively small.

3. The progress of theoretical study in green corrosion inhibitors

3.1 The behavior of green corrosion inhibitor studied by combination of quantum chemistry and QSAR

3.1.1 Carbon steel inhibitors

For carbon steel, the quantum chemical method is utilized to study and discuss the ability of ionic liquids to gain and lose electrons, reactive sites, intra-molecular and inter-molecular charge distribution, reactive parameters, and molar volume of the ionic liquid. The reaction parameters of protonation were calculated and the solvation effect was explained. Correlation equation between efficiency of inhibition corrosion and molecular structure parameters was established by QSAR to predict the performance of corrosion inhibitor on carbon steel surface [85-91].

The behavior of four imidazolium-based bis (trifluoromethyl-sulfonyl) imide ionic liquids inhibitors such as 1-propyl-3-methylimidazolium bis(trifluoromethyl-sulfonyl) imide [PMIM]NTf$_2$, 1-butyl-3-methylimidazolium bis(trifluoromethyl-sulfonyl) imide ([BMIM]NTf$_2$), 1-hexyl-3-methylimidazolium bis(trifluoromethyl-sulfonyl) imide ([HMIM]NTf$_2$) and 1-propyl-2,3-methylimidazolium bis(trifluoromethyl-sulfonyl) imide ([PDMIM]NTf$_2$) on carbon steel surface were studied by quantum chemical methods, weight loss and electrochemical measurements [85] in hydrochloric acid. It was found that ([PDMIM]NTf$_2$) has the strongest electron donating ability in four ionic liquids. Nucleophilic attack sites are all C2 atoms in the ring of the imidazole cation, the electrophilic points are oxygen and nitrogen atoms of the anion of ionic liquids. The cations are protonated and adsorbed on the metal surface. A good correlation between experimental efficiency of corrosion inhibitors and their molecular structure parameters obtained from quantum chemical calculation was established. It was found that [PDMIM]NTf$_2$ has largest inhibition efficiency, and [PMIM]NTf$_2$ has the smallest inhibition efficiency.

Mashuga et al. [25] studied the 1-hexyl-3-methylimidazolium based ionic liquids as green corrosion inhibitors such as 1-hexyl-3-methylimidazolium tetrafluoroborate ([HMIM]BF$_4$), 1-hexyl-3-methyl-imidazolium trifluoromethanesulfonate ([HMIM]TfO), 1-hexyl-3-methyl-imidazolium hexafluorophosphate ([HMIM]PF$_6$), and 1-hexyl-3-methyl-imidazolium iodide ([HMIM]I) for mild steel in 1 M HCl media with quantum chemical calculation and QSAR. Quantum chemical parameters obtained indicate that [HMIM]TfO and [HMIM]I have stronger electron supply capacity and reactivity. Anions of ionic liquids are the main active and adsorption sites on the surface. Anion with high electron density can transfer its electrons to carbon steel surface to bond with the surface. The order of corrosion inhibition efficiency is PF$_6^-$ < BF$_4^-$ < I$^-$ < TfO$^-$.

A QSAR model between experimental efficiency of corrosion inhibitors and their molecular structure parameters obtained from quantum chemical calculation was established by regression analysis for 20 benzimidazole derivatives [86]. Different benzene ring positions and groups have different electron-withdrawing abilities, and the adsorption energy on the metal surface is very different. The stronger the hydrogen bond is the higher inhibition efficiency is. From the QSAR model, it can be seen that the total charge has a good correlation with the inhibition efficiency. The longer the alkyl chain on benzimidazole is, the greater the electron density on the ring is, and the better the adsorption performance is.

The inhibitor behavior of six ionic liquids such as 1-butyl-3-methylimidazolium chloride ([BMIM]Cl), 1-ethyl-3-methyl-imidazolium iodide([EMIM]I), 1-butyl-3- methyl-imidazolium tetrafluoroborate ([BMIM]BF$_4$), 1-butyl-3-methylimidazolium hexafluoro-phosphate ([BMIM]PF$_6$),1-hexyl-3- methylimidazolium chloride ([HMIM]Cl), and 1-butyl-3- methylimidazolium bromide ([BMIM]Br) was studied [87]. The mixtures of these ionic liquids with sodium dodecyl sulfate (SDS) also studied and discussed. It was shown that comlex hydrogen-bonds were formed with imidazolium rings of ionic liquids and their counter ions, which has a great influence on the corrosion behavior of mild steel. A QSAR relationship between experimental efficiency of corrosion inhibitors and their molecular structure parameters obtained from quantum chemical calculation were established. It was found that the imidazole ring is the main reaction site, providing electrons and forming coordination bonds with the metal d orbitals. The order of corrosion inhibition is [BMIM]Cl < [BMIM]Br < [BMIM]BF$_4$ < [BMIM]PF$_6$.

The inhibition performance of six isomers of quinoxalinyl-dihydropyrazolyl -phenyl-propionyl-methanesulfonamide, such as N-(2-(1-propanoyl-5-(quinoxalin- 6-yl)-4,5-dihydro-1H-pyrazol-3-yl)phenyl) methanesulfonamide (MS-2-PQPP), N-(4- (1-(methanelsulfonyl)-5-(quinoxalin- 6-yl)-4,5-dihydro-1H- pyrazol-3-yl)phenyl) methane-sulfonamide(MS-4-PQPMS), N-(3-(1-propanoyl-5-(quinoxalin-6-yl)-4,5- dihydro-1H-pyrazol-3-yl)phenyl)methane-sulfon-amide (MS-3-PQPP), N-{3-[1- (methylsulfonyl)-5-(quinoxalin-6-yl)- 4,5-dihydro-1H-pyrazol-3-yl]phenyl} methane- sulfonamide (MS-3-PQPMS), N-(4-(1-propanoyl-5-(quinoxalin-6-yl)-4,5- dihydropyrazol-3-yl)phenyl) methanesulfonamide(MS-4-PQPP), and N-(2-(1- (methane-sulfonyl)-5-(quinoxalin-6-yl)-4,5-dihydro-1H-pyrazol-3-yl)phenyl) methanesulfonamide(MS-2-PQPMS) on carbon steel surface was investigated by theoretical methods [88]. There is a good correlation between experimental efficiency of corrosion inhibitors and their molecular structure parameters obtained from quantum chemical calculation. Moreover, the correlation between the protonated descriptor and the corrosion inhibition efficiency is better adsorbed on the metal surface through the pyrazole ring. The order of corrosion inhibition efficiency is MS-4-PQPMS < MS-2-PQPMS < MS-3-PQPMS, and MS-2-PQPP < MS-4-PQPP < MS-3-PQPP.

We used three types of ionic liquids, 1-alkyl-3-methylimidazole chloride ([XMIM]Cl), 1-alkyl-3-methylimidazolium acetate ([XMIM]Ac), 1-octyl-3-methyl- imidazole salt ([OMIM]Y), (X = ethyl, butyl, hexyl, octyl, and Y = Cl, BF$_4$, HSO$_4$, Ac, TfO) as corrosion inhibitor medium for corrosion protection of carbon steel [92,93]. It is found that with an increase the length of alkyl chain in the [XMIM]Cl and [XMIM]Ac systems, E_{HOMO}, E_{LUMO}, the softness, and polarizability increase gradually, whereas energy gap (ΔE), electronegativity, dipole moment, electrophilic index and hardness gradually decrease. For the [OMIM]Y system, with increaseing length of alkyl chain, the quantum chemical parameters of ionic liquids are quite different, and only the polarizability gradually decreases. The results show that inhibition is mainly [XMIM]$^+$ cations of the [XMIM]Cl system, and the order of inhibition efficiency follows as [OMIM]Cl > [HMIM]Cl > [BMIM]Cl > [EMIM]Cl. Both [XMIM]$^+$ cations and the Ac$^-$ anion have inhibition effect for the [XMIM]Ac system, and the order of inhibition efficiency is [OMIM]Ac > [HMIM]Ac> [BMIM]Ac > [EMIM]Ac. For the [OMIM]Y system, [XMIM]$^+$ cations and anions (BF$_4^-$, HSO$_4^-$, Ac$^-$, TfO$^-$) have inhibition effect, and the order of inhibition efficiency is [OMIM]TfO > [OMIM]Ac > [OMIM]HSO$_4$ > [OMIM]BF$_4$ > [OMIM]Cl. No matter what ionic liquids are used, cations or anions and cations of ionic liquids are adsorbed to cover on the Fe(110) surface as a protective film, which hinders the contact between iron surface and corrosion medium (H$_2$O, H$_3$O$^+$, Cl$^-$). In summary, the best corrosion inhibitor in the eleven studied ionic liquids is [OMIM]TfO.

It was found that the introduction of different groups such as NH$_2$, -CONH$_2$, -CH$_3$ and the ability of electron-withdrawing and electron-donating of long alkyl chains are unique, and their electrostatic interactions with copper surfaces are different. Substituents that have a tendency to form hydrogen bonds generally have strong electron-withdrawing capabilities. A good correlation between the total charge and sustained release efficiency was obtained through QSAR model analysis [85-92]. Theoretical calculations show that solitary pairs of inhibitors containing heteroatoms such as nitrogen, oxygen, and sulfur easily combine with the empty d orbitals on iron ions, which increases the adsorption energy. The corrosion inhibitor is easy to adsorb on the surface of metal, which prevents the diffusion of corrosion ions to contact with the metal surface, therefore the inhibitor effect is good.

We studied the mechanisms and behavior of ionic liquid corrosion inhibitors named 3-Decyl-1-methyl-imidazolium tetrafluoroborate ([C$_{10}$MIM]BF$_4$), 3-Dodecyl-1-methyl-imidazolium tetrafluoroborate ([C$_{12}$MIM]BF$_4$), 3-Decyl-1,2-dimethyl imidazolium tetra-fluoroborate ([C$_{10}$DMIM]BF$_4$) and 3-Dodecyl-1,2-dimethyl-imidazolium tetra-fluoroborate ([C$_{12}$DMIM]BF$_4$) on carbon steel using quantum chemistry and QSAR methods [94,95]. The imidazolium ring is mainly reactive site. With the increasing the alkyl chain length or the introduction of the methyl substituent on imidazolium ring, the electrons donor role of imidazolium ring, molecular reactivity, the interaction between corrosion inhibitors and the surface of carbon steel and the ability of diffusion inhibition for corrosive media particles increase gradually. The order of corrosion inhibition efficiency predicted by QSAR between inhibition efficiency and structural parameters is [C$_{12}$DMIM]BF$_4$ > [C$_{10}$DMIM]BF$_4$ > [C$_{12}$MIM]BF$_4$ > [C$_{10}$MIM]BF$_4$ which is consistent with the experimental result. The interaction between nitrogen atoms and Fe surface is stronger than carbon atoms on the imidazolium ring. Nitrogen atoms can donate electrons to Fe^{3+} or Fe^{2+} to form coordinate bonds.

3.1.2 Copper inhibitors

On the surface of copper, the correlation relationship between the experimental efficiency of inhibitors and their molecular structure parameters is studied through quantum chemistry and QSAR for thiophene, quinoxaline, benzothiol and its derivatives as corrosion inhibitors [96-98]. The QSAR method is applied in the preparation of a non-linear relationship between experimental inhibition efficiency of corrosion inhibitors and their molecular structure parameters. The principal reaction sites of the corrosion inhibitor can be obtained through quantum chemical calculations. The higher the electron density, the stronger the reactivity is. The reaction parameters can be utilized to clarify the reactivity trend of each atom. The electronic π structures of heteroatoms with lone and electron pairs, bicyclic, tricyclic, and aromatic rings can serve as adsorption centers, and they tend to supply electrons to metal surface [96-98].

The corrosion inhibition performance of 2,5-bis (3-thienyl)-4-amino-1,2,4-triazole (3-TAT) ionic liquids and 2,5-bis (2-thienyl) -4-amino- 1,2,4- triazole (2-TAT) ionic liquids on copper in 0.5M H_2SO_4 was studied [96]. The ability of two molecules to donate and accept electrons is similar. The reaction site of 2-TAT may be a triazole ring and a thiophene ring, while the 3-TAT triazole ring has a higher electron density, stronger reactivity, and corrosion inhibition higher efficiency.

The density functional theory calculation was performed on the B3LYP/LANL1MB theoretical level, and the clustering model was utilized to obtain binding energy of benzothiophene and its derivatives with the copper surface [97]. The clustering model and QSAR method are used to reveal inhibition behavior of corrosion inhibitors. Interactions between the inhibitors and metal surface occur mainly through sulfur and oxygen atoms.

The performance of quinoxaline derivatives as corrosion inhibitors for copper namely 2, 3-diphenyl quinoxaline (PPQ), 3-phenyl-2 (2-hydroxyphenyl) quinoxaline (PHPQ), 3-methyl-2-phenyl quinoxaline (MPQ), 3-methyl-2 (3-methoxy, 4-hydroxyphenyl) quinoxaline (MMtHPQ), and 3-methyl-2 2-hydroxyphenyl) quinoxaline (MHPQ) was studied by DFT and QSAR methods in acidic medium [98]. It was noted that C= C and nitrogen atoms are more inclined to accept electrons, while oxygen atoms are more inclined to provide lone pair electrons to the metal surface. Nitrogen atoms are readily protonated in acidic media. Physical adsorption is the primary factor, the higher the reactivity. The order of inhibition is MPQ < PPQ < MHPQ < PHPQ < MMtHPQ5.

We studied the corrosion inhibition mechanisms of 1-butyl-3-methyl-imidazolium hydrogen sulfate ([BMIM]HSO₄⁾, 1-octyl-3-methyl-imidazolium hydrogen sulfate ([OMIM]HSO₄), 1-hexyl-3-methyl -imidazolium hydrogen sulfate ([HMIM]HSO₄) on the Cu surface [94,95]. It was shown that the molecular reactivity, electrostatic interaction between corrosion inhibitors and the surface of copper and the ability of diffusion inhibition for corrosive media particles increase with the alkyl chain on imidazolium ring increasing. The order of corrosion inhibition efficiency predicted by QSAR between inhibition efficiencies and structural parameters is [BMIM]HSO$_4$ < [HMIM]HSO$_4$ < [OMIM]HSO$_4$, which agrees with the experimental results. The interaction between Carbon atoms and copper surface is stronger than nitrogen atoms on imidazolium ring. Corrosion inhibitors are adsorbed on Cu surface as protective film through physical adsorption, which cause the decrease of diffusion coefficients and number density for corrosive medium particles. The corrosion inhibition efficiency and concentration predicted by QSAR between inhibition efficiency and structural parameters are consistent with the experimental values, indicating that QSAR models established are reliable. The corrosion inhibitor can inhibit the contact between the corrosive medium and the metal surface effectively, which inhibits the electrochemical reaction and the ionization process of the metal.

3.2 The performance of green corrosion inhibitor studied by combination of molecular simulation and quantum chemistry

3.2.1 Carbon steel inhibitor

In recent years, many researchers have started to use molecular simulation to reveal the inhibition behavior of corrosion inhibitors on the surfaces of metal. Quantum chemistry and molecular simulation are powerful tools to elucidate the molecular characteristics and corrosion inhibition capabilities of organic inhibitors. These two techniques have been extensively used in the corrosion protection fields [99-108]. The corrosion inhibition of six different 1- (2-hydroxyethyl) -2-alkyl-imidazolines with different alkyl chain lengths on carbon steel surface was investigated [99]. It was noted that the N-C= N group of the reactive center is on the imidazoline. When carbon number of alkyl chain is greater than 13, a dense hydrophobic film is covered on the metal surface. The longer the alkyl chain, the higher the corrosion inhibition efficiency is.

Behavior of 1-butyl-3-methyl-1H-benzimidazole iodine ([BMBIM]I) inhibitor was studied by molecular simulation methods and quantum chemistry [100]. It finds that the benzimidazole ring is the active site for adsorption. Since benzimidazole is a planar conjugate structure, it is adsorbed in parallel on low carbon steel surface. The benzene ring can provide electrons to low carbon steel surface and form coordination bonds, while the π orbital of the imidazole ring can form a feedback bond with metal surface, there for a corrosion protective layer is formed.

19

The corrosion inhibition performance of three ionic liquids [BDMIM]BF$_4$ [EMIM]BF$_4$, and [C$_{10}$MIM]BF$_4$ with different chain lengths and substituents on carbon steel surface was investigated [101]. [C$_{10}$MIM]BF$_4$ has the longest alkyl chain, high reactivity, and strongest adsorption. Isolate the water molecules on the iron surface and form a durable protective layer to protect the iron from water corrosion. The introduction of the methyl group on the cation and the extension of the carbon chain enhance the electron density on the imidazole ring. The order of corrosion inhibition efficiency is [EMIM]BF$_4$ < [BDMIM]BF$_4$ < [C$_{10}$MIM]BF$_4$.

The corrosion inhibition mechanism of twenty N-alkyl-3-methylimidazolium bromides was studied [102]. It is found that the efficiency of corrosion inhibitor increases with the increase of the alkyl chain length. The IL with 18 carbon atoms on the branch chain is the most reactive corrosion inhibitor molecule with highest corrosion inhibition efficiency. For this ILs, the nucleophilic attack is on the imidazole ring of ionic liquids, while the electrophilic attack is on the alkyl chain of liquids.

Corrosion inhibition mechanism of1-vinyl-3-hexylimidazole iodide ([VHIM]I) in 0.5M sulfuric acid solution for low carbon steel was studied [103]. It was found that the adsorption process is mainly chemical adsorption. Nitrogen atoms in the imidazole ring can donate the electrons to carbon steel surface to form a protective film, which is adsorbed on the carbon steel surface in parallel and replaced the water molecules on the surface of iron.

The inhibition performance of 2-aminobenzothiazole (ABT) and 2-mercapto-benzothiazole (MBThione) is studied [104]. It was noted that the stability of the two molecules in solution was determined by the proton affinity. The nitrogen atom of ABT has the greatest negative electrostatic potential and is easily protonated. MBThione's sulfur and nitrogen atoms have high electron density, excessive electrons are supplied to the metal surface, and the benzene ring is the primary adsorption site, therefore MBThione has higher corrosion inhibition efficiency.

The performance of four pyrimidine derivatives ionic liquids corrosion inhibitor on carbon steel surface was investigated [105]. It was found that anions and cations contain heterocyclic pyrimidine derivatives provide the largest coverage area on the metal surface and have the highest inhibition efficiency. Heterocyclic electron donor ability is stronger than the alkyl chain and amino group, and it is more reactive. The adsorption of ILs on the surface of metal is chemisorption.

Quantum chemistry can reflect the electronic structure and geometric structure characteristics of molecules, and molecular dynamics simulation can be used to describe the inhibitor behavior of corrosion inhibitor molecules on low carbon steel surface in acidic solutions, and to reveal the behavior and mechanism of adsorption [99-108]. Most people further explain the experimental results by studying the adsorption energy and the orientation of the inhibitor molecules on surface of carbon steel. Generally, with increasing the alkyl chain of the cation, the adsorption energy and the area covered by the molecule on carbon steel surface is larger, the formed corrosion inhibitor film is more stable [99-101]. Inhibitor molecules replace water molecules on metal surface to form a durable protective layer which effectively drive away the corrosive media on the surface of metal [102].

Quantum chemical results show that the ability of electron donating of the heterocyclic ring is stronger than that of the alkyl chain and the amino group, and the reactivity is stronger. Most of the nitrogen, oxygen, and sulfur atoms on the imidazole ring donate lone pair electrons to the empty d orbital of the iron ion, while the π orbital on the benzene ring has a tendency to accept electrons from the iron atom. The charge transfer between metal surface and the molecule, most of the adsorption sites are located in the NC=N group on the imidazoline, which agrees well with quantum chemical calculations results [102-103]. Quantum chemical methods are widely applied to obtain the molecular structure and reactivity parameters of the inhibitors. The reactivity parameters reflect the adsorption trend. Molecular dynamics simulation can obtain the interaction between the inhibitors with the surface of carbon steel and the adsorption mechanism. But up to now, there is not a general formula about the correlation between molecular structure parameters and experimental inhibition efficiency, which is applicable to all systems. Again, for corrosion inhibitors with similar structure, it is impossible to predict the inhibition efficiency directly from the information of molecular structure parameters due to similarity principle.

3.2.2 Aluminum inhibitors

Some scholars have combined MD simulation and quantum chemistry to study the behavior of pyridine-type ionic liquids corrosion inhibitor and acetamide derivatives on aluminum surfaces in acidic solutions [106-107]. The results show that the reaction area of ionic liquids is located on heterocyclic rings. The electron density on the ring increases with introducing the methyl and methoxy groups. The increase in dipole moment indicates that its solvation effect is enhanced, and the dipole-dipole interaction in solution is enhanced. The solvation effect can change the solution The configuration of the inhibitor has important influence on the efficiency of corrosion inhibitor, and the root mean square displacement of the molecular dynamics simulation reflects that the three ionic liquids can inhibit the diffusion of the corrosive medium [106,107].

Theory and Applications of Green Corrosion Inhibitors   Materials Research Forum LLC
Materials Research Foundations **86** (2021) 1-37   https://doi.org/10.21741/9781644901052-1

The corrosion inhibition properties of three pyridine-type ionic liquids such as Pyridine, 3-methylpyridine, 4-methylpyridine and aluminum surfaces were examined in acidic solutions [106]. After the methyl group is introduced into the pyridine ring, the methyl group will have a conjugation effect with the nitrogen atom in the pyridine ring, thereby increasing the electron density on the ring, the interaction energy between the corrosion inhibitor and the copper surface will increase, and the carbon and nitrogen atoms will easily move Electrons are provided on the surface, and there is chemisorption between nitrogen atoms and the metal surface. The order of corrosion inhibition is pyridine < 3-methylpyridine < 4-methylpyridine.

For aluminum in alkaline environment, MC simulation, DFT methods and MD simulation were used to study the inhibition performance of N-Thiazolyl-2-cyanoacetamide derivatives [107]. It shows that the Al (111) with low surface energy has various adsorption sites. Therefore, it is suitable for the substrate of absorption. The reactive sites are located in the thiazole or benzene ring in three derivatives studied. The self-diffusion and the free volume are the two important factors that influence the inhibition behavior of the corrosion particle diffusion through the corrosion inhibitor film.

3.2.3 Copper inhibitors

Monte Carlo simulation can further analyze the corrosion inhibition performance of amino acids on the surface of copper in H_2O solution. Amino acids containing sulfur atoms have a strong electron-donating ability and higher corrosion inhibition efficiency [108]. Molecular dynamics simulation of four kinds of ionic liquids corrosion inhibitors such as 1-alkyl-3-butylimidazolium bromide([ABIM]Br), 1-alkyl-3-ethyl- imidazolium bromide ([AEIM]Br), 1-alkyl-3-octylimidazolium bromide ([AOIM]Br), 1-alkyl-3-hexylimidazolium bromide ([AHIM]Br) for copper in 0.5 M H_2SO_4 solution was performed [108]. It was shown that inhibitors with longer alkyl chain have higher adsorption, and a strong interaction between the corrosion inhibitor molecules and the metal surface. A feedback bond was formed by the π electrons in the imidazole ring and the copper surface. The longer the alkyl chain, the more area it will cover to metal surface. The alkyl chain is a hydrophobic branch, which narrows the contact of the corrosive medium. The corrosion inhibition properties of seven amino acids such as alanine, lysine, methionine, aspartic acid, asparagine, arginine and histidine on copper surface were studied [109]. Arginine, lysine and histidine contain more electron donating groups. It finds that the sulfur atom of methionine has a strong electron-donating ability. The interaction energy of inhibitor molecules on the surface of copper is much greater than that of water molecules, and gradually replaces water molecules from the copper surface to form a stable layer. The efficiencies of corrosion inhibitor increase with increasing the length of alkyl chain attached to imidazolium ring.

The performance of Thioflavin T (TT) and 3, 3-Diethylthiadicarbocyanine iodide (DI) inhibitors for copper in 0.1 M HCl solution was investigated [47]. It was found that the plane of the benzene ring is the active reaction center, which is mainly concentrated on olefinic double bonds and heteroatoms, indicating that this position is the electron-donating active center. The DI cation is the most likely to donate electrons. TT is absorbed by the entire molecular plane in a parallel mode. Due to the large steric hindrance of DI, the adsorption energy of TT is greater than DI and the corrosion inhibition efficiency is higher than DI.

Quantum chemistry and molecular simulation can effectively analyze the reactive sites of inhibitors, obtain the activity of each polar molecule, accept and donate electron capacity and adsorption tendency, explain how the inhibitor molecules form bonds with metal surface atoms, and judge adsorption mechanism and behavior. It is an effective method to reveal that how the inhibitor molecules form a protective film on the surface and inhibits the diffusion of corrosion media. However, the general correlation between the molecular structure parameter and experimental efficiency of corrosion inhibitor cannot be effectively analyzed for all corrosion inhibitors, so it is impossible to predict the effect of comparable inhibitor molecules on metal surface directly.

3.3 The behavior of green corrosion inhibitor studied by combination of molecular simulation, quantum chemistry and QSAR

The combination of molecular simulation, quantum chemistry and QSAR is wildly applied to study the behavior of ionic liquid as green inhibitors. The three methods are supplemented by each other. QSAR can be closely related to the inhibitory mechanism of inhibitor adsorption on metal surface. It can be applied to study the relationship between the multiple molecular structure parameters of inhibitor from quantum chemical calculation and the efficiency of corrosion inhibitor. Through quantum chemistry and molecular simulation, the interaction between the surfaces of metal and the corrosion inhibitor, the reactive sites, and the mechanism of the inhibitor can be analyzed to further verify the prediction of QSAR model [110-114]. It is not hard to find that at present, there is less work to study the corrosion inhibition performance of various metals by combining the three theoretical methods. In particular, there is the absence of research on QSAR, and there are few theoretical methods for the research on aluminum and copper corrosion inhibitor.

3.3.1 Carbon steel inhibitors

The relationship between the corrosion inhibitory efficiency of three corrosion inhibitors such as 4-chloro-acetophenone-oxy-1- (1,3,4-triazolyl) -toluoxime (CATM), 3,4-dichloro- acetophenone-oxy-1- (1,3,4-triazolyl) -acetoxime (DATM) and 4-fluoro-acetophenone- oxy-1- (1,3,4-tri (Oxazolyl) -toluoxime (FATM) and the molecular structure parameters from quantum chemical calculation and the adsorption configuration on the surface of iron were established and in gaseous and aqueous environments [110]. It was found that CATM and FATM are adsorbed on iron surface mainly through nitogen atoms, and DATM is adsorbed on the iron surface through nitrogen and chlorine atoms, which are parallel adsorption configurations. FATM holds the strongest electronic capability. In acidic solutions, molecules are available in protonated form. The order of the efficiency of corrosion inhibition is FATM > CATM > DATM.

The relationships between experimental efficiency of corrosion inhibitor and structural properties of 17 furan derivatives were established [111]. A good correlation between experimental efficiency of corrosion inhibitor and molecular structure parameters from quantum chemical calculations obtained. Among them, genetic function approximation (GFA) and neural network analysis (NNA) have high corrosion inhibition efficiency and the adsorption sites are located on the ring.

The behavior of five ionic liquids corrosion inhibitors such as 1-butyl-3-methyl -imidazole thiocyanate ([BMIM]SCN), 1-ethyl-3-methylimidazole acetate([EMIM]Ac), 1-butyl-3-methyl-imidazoleacetic acid ([BMIM]CH₃COO), 1-ethyl-3-methyl- imidazole ethyl sulfate ([EMIM]EtSO₄), and 1-butyl-3-methylimidazole dicyandiamide([BMIM]DCA) for low carbon steel and their mechanism under acidic environment were studied [112]. It was found that the nucleophilic attack area on ILs is located on the imidazole ring. The nitrogen, oxygen and sulfur atoms in the anion are more susceptible to electrophilic attack, forming a stable protective layer and replacing water molecules on iron surface. The order of efficiency of corrosion inhibitor is [EMIM]EtSO₄ > [BMIM]DCA > [BMIM]CH₃COO > [BMIM]SCN > [EMIM]Ac.

The performance of two kinds of ILs corrosion inhibitor such as 1- (6-ethoxy-6-oxohexyl) pyridine-1-bromide (S1) and 1- (2-bromoacetyl) pyridine bromide (S2) for carbon steel in hydrochloric acid was analyzed [113]. It was found that nitrogen, oxygen and sulfur atoms in two ionic liquids are the active centers to interact with the surface. The adsorption of pyridazine molecules on the surface of iron is mainly chemical adsorption. A coordination bond is formed by π electrons of the aromatic ring with the Fe surface. The order of corrosion inhibition efficiency is S1> S2.

MD simulations show that if the compounds studied are adsorbed in a planar or horizontal direction, chemical or physical adsorption occurs, which behaves as a good corrosion inhibitor, and the solvent effect makes the molecular structure and electronic structure of the corrosion inhibitor minor changes occur that increase the stability of the inhibitor film [110-113]. The cations are the same ones, and the effects of different anion types on corrosion inhibition are also very large. The anions containing sulfur, nitrogen, and oxygen atoms are provided by quantum chemical calculations with strong electron donating ability, strong adsorption, and carry methyl or heterocyclic. The imidazole cation has a strong electrostatic interaction with the surface [111]. The QSAR method shows that three to four quantum chemical parameters can establish a correlation between experimental and theoretical corrosion inhibition efficiency. Physical chemical description of interaction between iron surface and corrosion inhibitor were performed using genetic function approximation and neural network analysis technology. Correlation between inhibition efficiency, in which the physical and chemical descriptors consist of quantum chemical parameters, molecular surface area and volume, hydrophobicity coefficient, expansion coefficient, etc., and the molecular dynamics simulation analysis is used to study other green corrosion inhibitors such as furfur compounds [112]. It was found that furfur compounds are adsorbed on carbon steel surface through heteroatoms [112]. The molecular parameters obtained by quantum chemistry calculations include the energy levels of molecular orbits, dipole moments, solvation energy, Fukui index, average electron charge on nitrogen atoms were correlated to the inhibitor efficiency by QSAR method.

3.3.2 Copper inhibitors

The performance of five quinoxaline derivatives inhibitor molecules on the surface of copper in sulfuric acid solution was studied by Mwadham et al. [114]. A good QSAR correlation between molecular structure parameters from quantum chemical calculation and the efficiency of inhibition was obtained. It was found that there are multiple electron donors in the quinoxaline derivative, which enhances inhibition adsorption on the surface of metal. Metal atoms can simultaneously receive electrons from multiple donor centers. The formation of coordination bonds belongs to chemical adsorption. The efficiency of corrosion inhibitor varies with the position of substituent in quinoxaline derivatives.

Conclusions

The development and design of more green corrosion inhibitors with excellent inhibition performance, strong adaptability and significant economic benefits are particularly important for coping with various complex corrosion environments. Traditional corrosion inhibitor development methods are based on guesswork and a large number of experimental screenings. There are a series of problems such as high cost, long R & D cycle, and a certain degree of blindness in working. Therefore, the design and development of new green corrosion inhibitors urgently require theoretical guidance.

Although more and more corrosion inhibitor researchers have carried out various experimental explorations, the inhibition mechanism and behavior of corrosion inhibitor is still unclear. The theoretical research on the interaction between metal surface and corrosion inhibitors, adsorption mechanisms and micro-mechanisms is still insufficient, and the research work on the chemical and physical properties, the correlation between molecular structure parameters and corrosion inhibition efficiency is not detailed enough. Through theoretical calculations and molecular simulation, the mechanism of corrosion inhibitors can be revealed, which can provide guidance for the research and development of selecting and designing new and efficient corrosion inhibitors. Revealing the inhibition or absorption mechanism of green corrosion inhibitors using molecular simulation and quantum chemistry methods is a major research direction.

With the development of quantum chemistry theory and molecular simulation technology, the advancement of experimental technology and computer hardware level, it provides a new direction for the development and design of innovative green corrosion inhibitors. By constructing a quantitative structure-activity relationships (QSAR) model for corrosion inhibitors, it can provide guidance for designing, predicting, and synthesizing new and efficient green corrosion inhibitors. The use of molecular design methods to develop new type green inhibitors is one of the most important directions for corrosion inhibitor research. At the same time, making full use of relevant experimental results, improving the success rate of new green inhibitor molecular design, saving a lot of time and resources, reducing environmental pollution, and achieving the purpose of molecular tailoring inhibitors are also an important further research direction.

Design, development and synthesis of new highly effective green corrosion inhibitors play an important role. Although related research has made some progress, there are still many challenges to establish a truly effective approach to guide the progress of corrosion inhibitors through molecular design.

Acknowledgments

We would like to thank the grant support from National Natural Science Foundation of China (No.51774158, 51264021, 50904031) and Analysis and Testing Fund (2018M20172202025,2018M20172102005,2018M20172128007,2019M20182228005,2019M20182202030,2016M2014202009,2016M2014202059).

References

[1] M.M. Antonijevic, M.B. Petrovic. Copper corrosion inhibitors. A review. Int. J. Electroch. 3 (2008) 1-28.

[2] C. Verma, E.E. Ebenso, M.A. Quraishi. Ionic liquids as green and sustainable corrosion inhibitors for metals and alloys: an overview. J. Mol. Liq. 233 (2017) 403-414. https://doi.org/10.1016/j.molliq.2017.02.111

[3] D. Kesavan, M. Gopiraman, N. Sulochana. Green inhibitors for corrosion of metals: a review. Chem. Sci. Rev. Lett. 1(1) (2012) 1-8.

[4] K.Y. Quan. Research on the method and application of strong polarization detection of metal corrosion. Chongqing: Chongqing University, (2018) 8-12.

[5] J.H. Chen, S.D Zhang, X. Gao, J.R. Sun, P. Yan. Research status of temporary protection of steel under severe environmental conditions. Contemporary Chemical Industry, 48 (2019) 83-87.

[6] X. Wen, P. Bai, B. Luo, S. Zheng, C. Chen. Review of recent progress in the study of corrosion products of steels in a hydrogen sulphide environment, Corros. Sci. 139 (2018) 124-140. https://doi.org/10.1016/j.corsci.2018.05.002

[7] C. Verma, E.E. Ebenso, I. Bahadur, M.A. Quraishi. An overview on plant extracts as environmental sustainable and green corrosion inhibitors for metals and alloys in aggressive corrosive media. J. Mol. Liq. 266 (2018) 577-590.

[8] T.S. Zhang, H. Zhang, H. Gao, Inhibitors. Beijing: Chemical Industry Press, 2008, 316-339.

[9] W.H. Li. Synthesis and evaluation of new corrosion inhibitors, Beijing: Science Press, 2015, pp. 1-40.

[10] R.X. Li. Synthesis and application of green solvent-ionic liquid, Beijing: Chemical Engineering Press, 2004, pp. 2-10.

[11] Y. Deng. Ionic liquid-property, preparation and application. Beijing: China Petrochemical Press, 2006, pp. 10-23.

[12] S.J. Zhang, X.M. Lv, Z.P. Liu. Ionic liquids: from basic research to industrial applications . Beijing: Science Publishing, 2006, pp. 50-380.

[13] P.G. Yi, C.Z. Cao, X.Y.Wang. Synthesis of a novel ionic liquid and its cationic group corrosion inhibition performance. CIESC J., 56 (2005) 1112-1119.

[14] H. Ashassi-Sorkhabi, M. Es'haghi. Corrosion inhibition of mild steel in acidic media by [BMIM]Br Ionic liquid. Mater. Chem. Phys. 114 (2009) 267-271. https://doi.org/10.1016/j.matchemphys.2008.09.019

[15] Q.B. Zhang, Y.X. Hua. Corrosion inhibition of mild steel by alkylimidazolium ionic liquids in hydrochloric acid. Electrochim. Acta. 54 (2009) 1881-1887. https://doi.org/10.1016/j.electacta.2008.10.025

[16] S.K. Shukla, L.C. Murulana, E.E. Ebenso. Inhibitive effect of imidazolium based aprotic ionic liquids on mild steel corrosion in hydrochloric acid medium. Int. J. Electroch. 6 (2011) 4286-4295.

[17] X. Zhou, H. Yang, F. Wang. [BMIM]BF$_4$ ionic liquids as effective inhibitor for carbon steel in alkaline chloride solution. Electrochim. Acta. 56 (2011) 4268-4275. https://doi.org/10.1016/j.electacta.2011.01.081

[18] T. Tüken, F. Demir, N. Kıcır, G. Sığırcık, M. Erbil. Inhibition effect of 1-ethyl-3-methylimidazolium dicyanamide against steel corrosion. Corros. Sci. 59 (2012) 110-118. https://doi.org/10.1016/j.corsci.2012.02.021

[19] A. Zarrouk, M. Messali, H. Zarrok. Synthesis, characterization and comparative study of new functionalized imidazolium-based ionic liquids derivatives towards corrosion of C38 steel in molar hydrochloric acid. Int. J. Electrochem. Sci 7 (2012) 6998-7015.

[20] X. Zheng, S. Zhang, W. Li, L. Yin, J. He, J. Wu. Investigation of 1-butyl-3-methyl-1-H-benzimidazolium iodide as inhibitor for mild steel in sulfuric acid solution. Corros. Sci. 80 (2014) 383-392. https://doi.org/10.1016/j.corsci.2013.11.053

[21] E. Kowsari, M. Payami, R. Amini. Task-specific ionic liquid as a new green inhibitor of mild steel corrosion. Appl. Surf. Sci. 289 (2014) 478-486. https://doi.org/10.1016/j.apsusc.2013.11.017

[22] I. Lozano, E. Mazario, C.O. Olivares-Xometl. Corrosion behaviour of API 5LX52 steel in HCl and H$_2$SO$_4$ media in the presence of 1,3-dibencimidazolio acetate and 1, 3-dibencimidazolio dodecanoate ionic liquids as inhibitors. Mater. Chem. Phys. 147 (2014) 191-197. https://doi.org/10.1016/j.matchemphys.2014.04.029

[23] A. Acidi, M. Hasib-ur-Rahman, F. Larachi, A. Abbaci. Ionic liquids [EMIM]BF$_4$, [EMIM]OTf and [BMIM]OTf as corrosion inhibitors for CO$_2$ capture applications. Korean J. Chem. Eng. 31 (2014) 1043-1048. https://doi.org/10.1007/s11814-014-0025-3

[24] X. Zheng, S. Zhang, M. Gong, W. Li. Experimental and theoretical study on the corrosion inhibition of mild steel by 1-octyl-3-methylimidazolium l-prolinate in sulfuric

acid solution. Ind. Eng. Chem. Res. 53 (2014) 16349-16358.
https://doi.org/10.1021/ie502578q

[25] M.E. Mashuga, L.O. Olasunkanmi, A.S. Adekunle. Adsorption, thermodynamic and quantum chemical studies of 1-hexyl-3-methylimidazolium based ionic liquids as corrosion inhibitors for mild steel in HCl. Materials. 8 (2015) 3607-3632.
https://doi.org/10.3390/ma8063607

[26] Y. Ma, F. Han, Z. Li, C. Xia. Corrosion behavior of metallic materials in acidic-functionalized ionic liquids. ACS Sustain. Chem. Eng. 4 (2016) 633-639.
https://doi.org/10.1021/acssuschemeng.5b00974

[27] O. Olivares-Xometl, C. López-Aguilar, P. Herrastí-González, N. V. Likhanova, I. Lijanova, R. Martínez-Palou. Adsorption and corrosion inhibition performance by three new ionic liquids on API 5L X52 steel surface in acid media. Ind. Eng. Chem. Res. 53 (2014) 9534-9543. https://doi.org/10.1021/ie4035847

[28] Y. Ma, F. Han, Z. Li. Acidic-functionalized ionic liquid as corrosion inhibitor for 304 stainless steel in aqueous sulfuric acid. ACS Sustain. Chem. Eng. 4 (2016) 5046-5052. https://doi.org/10.1021/acssuschemeng.6b01492

[29] L. Feng, S. Zhang, S. Yan. Experimental and theoretical studies of 1-vinyl-3-hexylimidazolium iodide ([VHIM]I) as corrosion inhibitor for the mild steel in sulfuric acid solution. Int. J. Electroch. 12 (2017) 1915-1928.
https://doi.org/10.20964/2017.03.30

[30] C. Verma, I.B. Obot, I. Bahadur. Choline based ionic liquids as sustainable corrosion inhibitors on mild steel surface in acidic medium: Gravimetric, electrochemical, surface morphology, DFT and Monte Carlo simulation studies. Appl. Surf. Sci. 457 (2018) 134-149. https://doi.org/10.1016/j.apsusc.2018.06.035

[31] F. El-Hajjaji, M. Messali, A. Aljuhani. Pyridazinium-based ionic liquids as novel and green corrosion inhibitors of carbon steel in acid medium: electrochemical and molecular dynamics simulation studies. J. Mol. Liq. 249 (2018) 997-1008.
https://doi.org/10.1016/j.molliq.2017.11.111

[32] Y. Guo, Z. Chen, Y. Zuo. Ionic liquids with two typical hydrophobic anions as acidic corrosion inhibitors. J. Mol. Liq. 269 (2018) 886-895.
https://doi.org/10.1016/j.molliq.2018.08.090

[33] C. Verma, L.O. Olasunkanmi, I. Bahadur. Experimental, density functional theory and molecular dynamics supported adsorption behavior of environmental benign

imidazolium based ionic liquids on mild steel surface in acidic medium. J. Mol. Liq. 273 (2019) 1-15. https://doi.org/10.1016/j.molliq.2018.09.139

[34] S. Cao, D. Liu, H. Ding. Corrosion inhibition effects of a novel ionic liquid with and without potassium iodide for carbon steel in 0.5 M HCl solution: An experimental study and theoretical calculation. J. Mol. Liq. 275 (2019) 729-740. https://doi.org/10.1016/j.molliq.2018.11.115

[35] P. Arellanes-Lozada, O. Olivares-Xometl, N.V. Likhanova. Adsorption and performance of ammonium-based ionic liquids as corrosion inhibitors of steel. J. Mol. Liq. 265 (2018) 151-163. https://doi.org/10.1016/j.molliq.2018.04.153

[36] N.V. Likhanova, P. Arellanes-Lozada, O. Olivares-Xometl. Effect of organic anions on ionic liquids as corrosion inhibitors of steel in sulfuric acid solution. J. Mol. Liq. 279 (2019) 267-278. https://doi.org/10.1016/j.molliq.2019.01.126

[37] V. Branzoi, F. Golgovici, F. Branzoi. Aluminium corrosion in hydrochloric acid solutions and the effect of some organic inhibitors. Mater. Chem. Phys. 78 (2003) 122-131. https://doi.org/10.1016/S0254-0584(02)00222-5

[38] Q.B. Zhang, Y.X. Hua. Corrosion inhibition of aluminum in hydrochloric acid solution by alkylimidazolium ionic liquids. Mater. Chem. Phys. 119 (2010) 57-64. https://doi.org/10.1016/j.matchemphys.2009.07.035

[39] X. Li, S. Deng, H. Fu. Inhibition by tetradecylpyridinium bromide of the corrosion of aluminium in hydrochloric acid solution. Corros. Sci. 53 (2011) 1529-1536. https://doi.org/10.1016/j.corsci.2011.01.032

[40] P. Arellanes-Lozada, O. Olivares-Xometl, D. Guzmán-Lucero, N. Likhanova, M. Domínguez-Aguilar, I. Lijanova, E. Arce-Estrada. The inhibition of aluminum corrosion in sulfuric acid by poly (1-vinyl-3-alkyl- imidazolium hexafluorophosphate). Materials. 7 (2014) 5711-5734. https://doi.org/10.3390/ma7085711

[41] S.K. Shetty, A.N. Shetty. Ionic liquid as an effective corrosion inhibitor on 6061 Al-15 Vol. Pct. SiC(p) composite in 0.1 M H_2SO_4 medium–an ecofriendly approach. Can. Chem. Trans. 3 (2015) 41-64. https://doi.org/10.13179/canchemtrans.2015.03.01.0160

[42] P. Huang, A. Somers, P.C. Howlett. Film formation in trihexyl (tetradecyl) phosphonium diphenylphosphate ([P6,6,6,14]dpp) ionic liquid on AA5083 aluminium alloy. Surf. Coat. Tech. 303 (2016) 385-395. https://doi.org/10.1016/j.surfcoat.2015.12.060

[43] I. Perissi, U. Bardi, S. Caporali. High temperature corrosion properties of ionic liquids, Corros. Sci. 48 (2006) 2349-2362. https://doi.org/10.1016/j.corsci.2006.06.010

[44] Q.B. Zhang, Y.X. Hua. Effect of alkylimidazolium ionic liquids on the corrosion inhibition of copper in sulfuric acid solution. Acta Phys.-Chim. Sin. 27 (2011) 655-663.

[45] M. Cai, Y. Liang, F. Zhou, W. Liu. Anticorrosion imidazolium ionic liquids as the additive in poly (ethylene glycol) for steel/Cu–Sn alloy contacts, Faraday Discuss. 156 (2012) 147-157. https://doi.org/10.1039/c2fd00124a

[46] Y. Qiang, S. Zhang, L. Guo, X. Zheng, B. Xiang, S. Chen. Experimental and theoretical studies of four alkyl imidazolium-based ionic liquids as green inhibitors for copper corrosion in sulfuric acid, Corros. Sci. 119 (2017) 68-78. https://doi.org/10.1016/j.corsci.2017.02.021

[47] L. Feng, S. Zhang, Y. Lu. Synergistic corrosion inhibition effect of thiazolyl-based ionic liquids between anions and cations for copper in HCl solution, Appl. Surf. Sci., 483 (2019) 901-911. https://doi.org/10.1016/j.apsusc.2019.03.299

[48] C.W. Yan, H. C. Lin, C.N. Cao. Valence state and film formation mechanism of cu in MBO inhibition films, Corros. Sci. and Protec. Tech. 13 (2001) 12-15.

[49] H. Yu, J.H. Wu, J.H. Qian. Research on corrosion inhibition behavior of a seawater corrosion inhibitor, J. Chin. Soc. Corr. Pro., 23 (2003) 295-298.

[50] X.Q. Jie, Z.G. Ding, L. Zhu. SERs study of corrosion inhibition of btah and its derivative on copper electrode in sulphuric acid solution, J. Chin. Soc. Corr. Pro. 21 (2001) 172-176.

[51] K. Yu, C.C. Xu. Inhibition of sodium molybdate on iron cultural relics, J. B. Univ. Chem. Techol (Nat. Sci. Ed.). 31 (2004) 41-44.

[52] R.G. Parr and W.T. Yang. Density-functional theory of atoms and molecules: Oxford University Press, New York, and Clarendon Press, Oxford, 1989.

[53] J.F. Dobson, G. Vignale, M.P. Das, (Eds.). Electronic density functional theory: recent progress and new directions. Springer Science & Business Media, 2013.

[54] J.K. Labanowski, J.W. Andzelm (Eds.), Density Functional Methods in Chemistry, Springer Science & Business Media, 2012.

[55] J. Vosta, J. Eliasek, Study on corrosion inhibition from aspect of quantum chemistry, Corros. Sci. 11 (1971) 223-229. https://doi.org/10.1016/S0010-938X(71)80137-3

[56] T. Koopmans, Ordering of wave functions and eigenenergies to the individual electrons of an atom, Physica. 1 (1933) 104-113. https://doi.org/10.1016/S0031-8914(34)90011-2

[57] S. Kaya, C. Kaya, I.B. Obot, N. Islam, A novel method for the calculation of bond stretching force constants of diatomic molecules, Spectrochim. Acta. A. 154 (2016) 103-107. https://doi.org/10.1016/j.saa.2015.10.030

[58] H. Chermette, Chemical reactivity indexes in density functional theory, J. Comput. Chem. 20 (1999) 129-154. https://doi.org/10.1002/(SICI)1096-987X(19990115)20:1<129::AID-JCC13>3.0.CO;2-A

[59] R.G. Parr, P.K. Chattaraj, Principle of maximum hardness, J. Am. Chem. Soc. 113 (1991) 1854-1855. https://doi.org/10.1021/ja00005a072

[60] R.P. Iczkowski, J.L. Margrave, Electronegativity, J. Am. Chem. Soc. 83 (1961) 3547-3551. https://doi.org/10.1021/ja01478a001

[61] W. Yang, R.G. Parr, Hardness, softness, and the fukui function in the electronic theory of metals and catalysis, P. Natl. Acad. Sci. Usa. 82 (1985) 6723-6726. https://doi.org/10.1073/pnas.82.20.6723

[62] A.Y. Musa, A.A.H. Kadhum, A.B. Mohamad, A.A.B. Rahoma, H. Mesmari, Electrochemical and quantum chemical calculations on 4, 4-dimethyloxazolidine-2-thione as inhibitor for mild steel corrosion in hydrochloric acid, J. Mol. Struct. 969 (2010) 233-237. https://doi.org/10.1016/j.molstruc.2010.02.051

[63] M. Shahraki, M. Dehdab, S. Elmi, Theoretical studies on the corrosion inhibition performance of three amine derivatives on carbon steel: molecular dynamics simulation and density functional theory approaches, J. Taiwan Inst. Chem. E. 62 (2016) 313-321. https://doi.org/10.1016/j.jtice.2016.02.010

[64] R.G. Parr, W. Yang, Density functional approach to the frontier-electron theory of chemical reactivity, J. Am. Chem. Soc. 106 (1984) 4049-4050. https://doi.org/10.1021/ja00326a036

[65] K.F. Khaled, Experimental, density function theory calculations and molecular dynamics simulations to investigate the adsorption of some thiourea derivatives on iron surface in nitric acid solutions, Appl. Surf. Sci. 256 (2010) 6753-6763. https://doi.org/10.1016/j.apsusc.2010.04.085

[66] C. Verma, I.B. Obot, I. Bahadur, E.S.M. Sherif, E.E. Ebenso, Choline based ionic liquids as sustainable corrosion inhibitors on mild steel surface in acidic medium: Gravimetric, electrochemical, surface morphology, DFT and Monte Carlo Simulation studies, Appl. Surf. Sci. 457 (2018) 134-149. https://doi.org/10.1016/j.apsusc.2018.06.035

[67] S. Martinez, Inhibitory mechanism of mimosa tannin using molecular modeling and substitutional adsorption isotherms, Mater. Chem. Phys. 77 (2003) 97-102. https://doi.org/10.1016/S0254-0584(01)00569-7

[68] S. Kaya, C. Kaya, L. Guo, F. Kandemirli, B. Tüzün, İ. Uğurlu, M. Saraçoğlu, Quantum chemical and molecular dynamics simulation studies on inhibition performances of some thiazole and thiadiazole derivatives against corrosion of iron, J. Mol. Liq. 219 (2016) 497-504. https://doi.org/10.1016/j.molliq.2016.03.042

[69] B. Gómez, N.V. Likhanova, M.A D. Aguilar, O. Olivares, J.M. Hallen, J.M. Martínez-Magadán, Theoretical study of a new group of corrosion inhibitors. J. Phys. Chem. A., 2005, 109(39): 8950-8957. https://doi.org/10.1021/jp052188k

[70] K. Fukui, Role of frontier orbitals in chemical reactions, Science 218 (1982) 747-754. https://doi.org/10.1126/science.218.4574.747

[71] K.X. Chen, Computer-Aided Drug Design: Principles, Methods and Applications, Shanghai Science and Technology Press, 2000.

[72] X.J. Xu. Computer-aided drug molecular design Chemical Industry Press, 2004.

[73] I. Lukovits, I. Bakó, A. Shaban, E. Kálmán, Polynomial model of the inhibition mechanism of thiourea derivatives. Electrochim. Acta, 43 (1998) 131-136. https://doi.org/10.1016/S0013-4686(97)00241-7

[74] P.G. Abdul-Ahad, S.H.F Al-Madfai, Elucidation of corrosion inhibition mechanism by means of calculated electronic indexes, Corros. Sci., 45 (1989) 978-980. https://doi.org/10.5006/1.3585015

[75] F. Bentiss, M. Traisnel, H. Vezin, Linear resistance model of the inhibition mechanism of steel in HCl by triazole and oxadiazole derivatives: structure–activity correlations, Corros. Sci., 45 (2003) 371-380. https://doi.org/10.1016/S0010-938X(02)00102-6

[76] S.G. Zhang, W. Lei, M.Z. Xia, QSAR study on N-containing corrosion inhibitors: Quantum chemical approach assisted by topological index. J. Mol. Struc-THEOCHEM. 732 (2005) 173-182. https://doi.org/10.1016/j.theochem.2005.02.091

[77] E.S.H.E. Ashry, A.E. Nemr, S.A. Esawy, Corrosion inhibitors Part II: Quantum chemical studies on the corrosion inhibitions of steel in acidic medium by some triazole, oxadiazole and thiadiazole derivatives, Electrochim. Acta. 51 (2006) 3957-3968. https://doi.org/10.1016/j.electacta.2005.11.010

[78] E.S.H.E. Ashry, A.E. Nemr, S.A. Esawy, Corrosion inhibitors Part V: QSAR of benzimidazole and 2-substituted derivatives as corrosion inhibitors by using the quantum

chemical parameters, Prog. Org. Coat. 61 (2008) 11-20.
https://doi.org/10.1016/j.porgcoat.2007.08.009

[79] S.Q. Hu, J.C. Hu, X. Shi, Quantitative structure-activity relationship and molecular design of imidazoline derivatives inhibitors, Acta Phys. Sin. 25 (2009) 2524-2530. https://doi.org/10.3866/PKU.WHXB20091126

[80] S.K. Saha, P. Ghosh, A. Hens, Density functional theory and molecular dynamics simulation study on corrosion inhibition performance of mild steel by mercapto-quinoline Schiff base corrosion inhibitor, Physica. E. 66 (2015) 332-341.
https://doi.org/10.1016/j.physe.2014.10.035

[81] H. Zhao, X. Zhang L. Ji. Quantitative structure–activity relationship model for amino acids as corrosion inhibitors based on the support vector machine and molecular design, Corros. Sci. 83 (2014) 261-271. https://doi.org/10.1016/j.corsci.2014.02.023

[82] S. Ramachandran, B.-L. Tsai, M. Blanco, Atomistic simulations of oleic imidazolines bound to ferric clusters, J. Phys. Chem. A. 101(1997) 83-88.
https://doi.org/10.1021/jp962041g

[83] X.C. Yan, H. Zhao, M.D. Luo, Current status and prospects of quantum chemistry research on corrosion inhibition mechanism of metals, Chin. J. Nature, 1998 (3) 134-137.

[84] T.S. Lawrence, Y. Wei, S.A. Jansen, Corrosion inhibition by aniline oligomers through charge transfer: a DFT approach, Synth. Met. 143(2004) 1-12.
https://doi.org/10.1016/j.synthmet.2002.06.002

[85] L.C. Murulana, A.K. Singh, S.K. Shukla, M.M. Kabanda, E. E. Ebenso, Experimental and quantum chemical studies of some bis (trifluoromethyl-sulfonyl) imide imidazolium-based ionic liquids as corrosion inhibitors for mild steel in hydrochloric acid solution, Ind. Eng. Chem. Res. 51 (2012) 13282-13299.
https://doi.org/10.1021/ie300977d

[86] X.H. Du, C.J. Feng, Prediction of corrosion inhibition efficiency of benzimidazole inhibitors based on density functional theory, J. Nanjing U. Sci. Tech: Nat. Sci. Ed., 38 (2014) 424-430.

[87] A. Yousefi, S. Javadian, N. Dalir, J. Kakemam, J. Akbari, Imidazolium-based ionic liquids as modulators of corrosion inhibition of SDS on mild steel in hydrochloric acid solutions: experimental and theoretical studies, RSC Advances 5 (2015) 11697-11713.
https://doi.org/10.1039/C4RA10995C

[88] L.O. Olasunkanmi, I.B. Obot, E.E. Ebenso, Adsorption and corrosion inhibition properties of N-{n-[1-R-5-(quinoxalin-6-yl)-4,5-dihydropyrazol-3-yl]phenyl} methane-

Materials Research Forum LLC
https://doi.org/10.21741/9781644901052-1

sulfonamides on mild steel in 1 M HCl: experimental and theoretical studies, RSC Advances. 6 (2016) 86782-86797. https://doi.org/10.1039/C6RA11373G

[89] Y.M. Tang, Y. Chen, W.Z. Yang, Y. Liu, X.S. Yin, J.T. Wang, Electrochemical and theoretical studies of thienyl-substituted amino triazoles on corrosion inhibition of copper in 0.5 M H$_2$SO$_4$, J. Appl. Electrochem. 38 (2008) 1553-1559. https://doi.org/10.1007/s10800-008-9603-6

[90] M. Mousavi, T. Baghgoli, Application of interaction energy in quantitative structure-inhibition relationship study of some benzenethiol derivatives on copper corrosión, Corros. Sci., 105 (2016) 170-176. https://doi.org/10.1016/j.corsci.2016.01.014

[91] M.M. Kabanda, E.E. Ebenso, Density functional theory and quantitative structure-activity relationship studies of some quinoxaline derivatives as potential corrosion inhibitors for copper in acidic médium, Int. J. Electroch. 7(2012) 8713-8733.

[92] W.Z. Zhou, G.C. Tian, Theoretical study on corrosion inhibition mechanism of alkylimidazole ionic liquid corrosion inhibitors, J. Kunming Univ. Sci. Tech.(Nat. Sci. Ed.). 42 (2017) 1-9.

[93] W.Z. Zhou, Theoretical study on the inhibition mechanism of ionic liquid corrosion inhibitors, Kunming University of Science and Technology: Kunming, 2018

[94] G.C. Tian, K.T. Yuan, Mechanism of inhibition of carbon steel by imidazole ionic liquid in hydrochloric acid médium, J. Kunming Univ. Sci. Tech. (Nat. Sci. Ed.), 43 (2018) 9-19.

[95] Yuan Kaitao, Quantitative structure-activity relationship study of new ionic liquid corrosion inhibitors, Kunming University of Science and Technology: Kunming, 2019, 95-96.

[96] Y.M. Tang, Y. Chen, W.Z. Yang, Y. Liu, X.S. Yin, J.T. Wang, Electrochemical and theoretical studies of thienyl-substituted amino triazoles on corrosion inhibition of copper in 0.5 M H$_2$SO$_4$, J. Appl. Electrochem. 38 (2008) 1553-1559. https://doi.org/10.1007/s10800-008-9603-6

[97] M. Mousavi, T. Baghgoli, Application of interaction energy in quantitative structure-inhibition relationship study of some benzenethiol derivatives on copper corrosión, Corros. Sci. 105 (2016) 170-176. https://doi.org/10.1016/j.corsci.2016.01.014

[98] M.M. Kabanda, E.E. Ebenso, Density functional theory and quantitative structure-activity relationship studies of some quinoxaline derivatives as potential corrosion inhibitors for copper in acidic médium, Int. J. Electroch. 7 (2012) 8713-8733.

[99] J. Zhang, S.Q. Hu, Y. Wang, W.Y. Guo, J. X. Liu, L. You, Theoretical Study on the Inhibition Mechanism of 1- (2-hydroxyethyl) -2-alkyl-imidazoline Corrosion Inhibitors, Acta Chim. Sin. 66 (2008) 2469-2475.

[100] X. Zheng, S. Zhang, W. Li, L. Yin, J. He, J. Wu, Investigation of 1-butyl-3-methyl-1-H-benzimidazolium iodide as inhibitor for mild steel in sulfuric acid solution, Corros. Sci. 80 (2014) 383-392. https://doi.org/10.1016/j.corsci.2013.11.053

[101] Y. Sasikumar, A.S. Adekunle, L.O. Olasunkanmi, I. Bahadur, R. Baskar, M.M. Kabanda, E.E. Ebenso, Experimental, quantum chemical and Monte Carlo simulation studies on the corrosion inhibition of some alkyl imidazolium ionic liquids containing tetrafluoroborate anion on mild steel in acidic médium, J. Mol. Liq. 211(2015) 105-118. https://doi.org/10.1016/j.molliq.2015.06.052

[102] C. Zuriaga-Monroy, R. Oviedo-Roa, L.E. Montiel-Sánchez, A. Vega-Paz, J. Marín-Cruz, J.M. Martinez-Magadan, Theoretical study of the aliphatic-chain length's electronic effect on the corrosion inhibition activity of methylimidazole-based ionic liquids, Ind. Eng. Chem. Res. 55(2016) 3506-3516. https://doi.org/10.1021/acs.iecr.5b03884

[103] L. Feng, S. Zhang, S. Yan, S. Xu, S. Chen, Experimental and theoretical studies of 1-vinyl-3-hexylimidazolium iodide ([VHIM]I) as corrosion inhibitor for the mild steel in sulfuric acid solution. Int. J. Electroch., 12 (2017) 1915-1928. https://doi.org/10.20964/2017.03.30

[104] R.H. Albrakaty, N.A. Wazzan, I.B. Obot, Theoretical study of the mechanism of corrosion inhibition of carbon steel in acidic solution by 2-aminobenzothaizole and 2-mercatobenzothiazole, Int. J. Electroch., 13 (2018) 3535-3554. https://doi.org/10.20964/2018.04.50

[105] F.E.T Heakal, S.A. Rizk, A.E. Elkholy, Characterization of newly synthesized pyrimidine derivatives for corrosion inhibition as inferred from computational chemical análisis, J. Mol. Struc. 1152 (2018) 328-336. https://doi.org/10.1016/j.molstruc.2017.09.079

[106] G. Wu, N.M. Hao, B.J. Lian, S. Chen, S. Sun, S. Hu, Density functional theory analysis on pyridine corrosion inhibitors and adsorption behavior on Al (111) Surface, CIESC J. 64 (2013) 2565-2572.

[107] X.Y. Zhang, Q.X. Kang, Y. Wang, Theoretical study of N-thiazolyl-2 -cyanoacetamide derivatives as corrosion inhibitor for aluminum in alkaline environments, Comput. Theor. Chem. 1131(2018) 25-32. https://doi.org/10.1016/j.comptc.2018.03.026

[108] Y. Qiang, S. Zhang, L. Guo, Experimental and theoretical studies of four allyl imidazolium-based ionic liquids as green inhibitors for copper corrosion in sulfuric acid, Corros. Sci. 119 (2017) 68-78. https://doi.org/10.1016/j.corsci.2017.02.021

[109] S. Kaya, B. Tüzün, C. Kaya, I.B. Obot, Determination of corrosion inhibition effects of amino acids: Quantum chemical and molecular dynamic simulation study, J. Taiwan Inst. Chem. Eng. 58 (2016) 528-535. https://doi.org/10.1016/j.jtice.2015.06.009

[110] L. Guo, S. Zhu, S. Zhang, Q. He, W. Li, Theoretical studies of three triazole derivatives as corrosion inhibitors for mild steel in acidic médium, Corros. Sci. 87 (2014) 366-375. https://doi.org/10.1016/j.corsci.2014.06.040

[111] K.F. Khaled, N.M. Al-Nofai, N.S. Abdel-Shafi, QSAR of corrosion inhibitors by genetic function approximation, neural network and molecular dynamics simulation methods. J. Mater. Environ. Sci. 7(6) (2016) 2121-2136.

[112] S. Yesudass, L.O. Olasunkanmi, I. Bahadur, M.M. Kabanda, I. B. Obot, E.E. Ebenso, Experimental and theoretical studies on some selected ionic liquids with different cations/anions as corrosion inhibitors for mild steel in acidic médium, J. Taiwan Inst. Chem. Eng. 64 (2016) 252-268. https://doi.org/10.1016/j.jtice.2016.04.006

[113] F. El-Hajjaji, M. Messali, A. Aljuhani, M.R. Aouad, B. Hammouti, M.E. Belghiti, M.A. Quraishi, Pyridazinium-based ionic liquids as novel and green corrosion inhibitors of carbon steel in acid medium: Electrochemical and molecular dynamics simulation studies, J. Mol. Liq. 249 (2018) 997-1008. https://doi.org/10.1016/j.molliq.2017.11.111

[114] M.M. Kabanda, E.E. Ebenso, Density functional theory and quantitative structure-activity relationship studies of some quinoxaline derivatives as potential corrosion inhibitors for copper in acidic médium, Int. J. Electroch. 7 (2012) 8713-8733

Materials Research Forum LLC
https://doi.org/10.21741/9781644901052-2

Chapter 2

Effect of Natural Sources on the Corrosion Inhibition

M. Bugdayci

Yalova University, Chemical Engineering Department, Yalova, 77200, Turkey

Istanbul Medipol University, Vocational School, Construction Technology Dep., 34810, Istanbul, Turkey

mehmet.bugdayci@yalova.edu.tr

Abstract

Corrosion costs more than a trillion dollars worldwide and with great efforts being made to solve this problem. The toxic effect of components used in the solution of this problem causes new problems with increasing environmental concerns and high costs. Accordingly, green inhibitors have become an alternative protection method with its ecofriendly nature. With the development of green chemistry technologies, the role of plant extracts and biowastes as new corrosion inhibitors in various environments and their protection effect for many materials have been determined. In this study, besides these mentioned topics, test methods, corrosion prevention mechanisms, and economic approaches are summarized.

Keywords

Green Inhibitors, Corrosion Prevention, Natural Inhibitors, Environmental Science, Anodic Protection

Contents

1.   Introduction

Corrosion is the change or wear event of the material starting from the surface and creating an effect with a chemical and electrochemical reaction [1]. Corrosion is the opposite of the production process of metals. Almost all of the technologically important metals, except noble metals, are present in nature as "compounds" [2]. The industry normally produces pure metals and alloys, but there is always a tendency to return to compound form. As a result of this, metallic materials react with the elements of the medium in which they are, first they become ionic and then they become compounded with other elements in the environment, they undergo chemical change and decompose [3]. After all; the chemical, physical or mechanical properties of metals or alloys undergo undesirable changes. Corrosion refers to both the decomposition reaction of metal and alloy (oxidation) and the damage caused by this reaction [4].

All metal structures undergo some degree of corrosion in the natural environment [5]. Bronze, brass, stainless steel, zinc and aluminum materials can last longer without protection [6]. Structural corrosion of iron and steel progresses rapidly if the metal is not protected to the extent necessary [7]. This corrosive sensitivity of iron and steel is an important focus of interest [8]. Because of the great costs and physical properties are used in large quantities. The protection of iron and steel against corrosion is of great importance to maintenance engineers [9].

In order to limit the negative effects of corrosion on materials, the surfaces are kept away from the aggressive environment by coating or film forming methods. Corrosion measures are also taken by controlling the electron cycle as a result of cathodic protection of metals or by using another material added as an anode to the structure [1].

Corrosion limitation is the process of isolating the reduction process in metals with organic or inorganic compounds. Organic inhibitors are absorbed into the surface exposed to aggressive environment and the formed layer is capable of protecting the surface [10].

As an alternative to organic inhibitors, inorganic inhibitors provide anodic protection and add metal atoms to the film to increase corrosion resistance. Most of the researchers have seen the toxic effects of corrosion inhibitors and stated that recycling processes cause serious environmental hazards. The use of toxic anode materials is limited due to laws enacted by environmental concerns. Therefore, taking environmentally friendly corrosion measures from natural sources has become an important requirements [11,12].

Electrolytic corrosion inhibitors; It is divided into two as anodic and cathodic inhibitors. In addition, it is formed under the conditions where the two types of inhibitors are used together in corrosion measures performed according to the working conditions of the material.

The use of hazardous chemicals and the elimination of their toxic effects have emerged as a popular issue among researchers in the last decade. Therefore, new chemical technologies have been developed and commercialized to prevent toxin effect of processes and reduce waste [13]. These concerns that emerged in the 1980s led to the adoption of a US anti-pollution law in 1990. A few years later, "12 Principles of Green Chemistry" was announced by Anastas for solutions to environmental problems [14]. After that, studies from the UK to Japan became widespread worldwide. Over the years, it has been concluded that the basic scientific methodologies of green chemistry can economically protect human health and the environment. With the understanding of the importance of corrosion, many research & development, education and social responsibility projects related to the subject have been adapted to this topic. These researches are carried out by designing polymers [15], solvents [16], catalysis [17], analytical method and synthetic method applications [18,19] and harmless environmentally friendly chemical uses.

Sustainability principles in green chemistry and industrial applications are provided by biomass and renewable raw materials. The fact that chemical compounds found in nature offer alternative compounds in high combinations increases the interest in renewable raw materials. Ensuring the ecological balance of living organisms will be possible with the use of these substances [20–22]. One of the important research areas of green chemistry is the protection of metallic surfaces from chemical degradation. The applications for the control of metallic corrosion were determined as alloying, coating and correct system design. In addition, inhibitors are an important alternative to these anti-corrosion methods [23]. Green chemistry, using harmless preservatives, is of great interest in the area of corrosion protection because of its safe, ecological, biodegradable, renewable and sustainability properties [24]. These natural preservatives have been identified as plant extracts, animal products, amino acids, alkaloids, by-products of agricultural waste and polyphenols [22,25–27].

Green inhibitors are important components used in aggressive media for the corrosion limitation of ferrous and non-ferrous materials and alloys. Plant extracts are naturally occurring chemical compounds which, when added to metal surfaces, significantly reduce the corrosion rate by diffusion of suitable components [28]. These corrosion inhibitors obtained from natural sources; it creates a protective film on the surface by manipulating the corrosion reactions and diffusion mechanisms occurring between the aggressive media and the surface, this film performs its function by preventing electron transfer by reducing the current permeability on the surface. The mechanism of decomposition is based on the transfer of ionizing metal electrons from the anode to the cathode. The ionization of the metal requires the presence of oxygen, oxidizing material or hydrogen ions, and these factors are largely present where metallic structures are present. Delaying or preventing these reactions from occurring completely stops or minimizes corrosion. Inhibitors prevent the oxidation by absorbing to the surface, forming inhibition film, and interacting with aggressive media-metal reactions [29].

In cathodic protection, it is essential to stop the interaction of hydrogen ions with the metal surface. The protective film layer which is made up of inhibitors, prevents the penetration of hydrogen ions into the protected surface, preventing corrosion by substituting neutral compounds instead. The green inhibitors have adsorptive properties that are capable of blocking the aggressive medium [30,31]. Research in this area has focused on the use of harmless and natural compounds, which are defined as green corrosion inhibitors to protect different alloys and metals against corrosion.

2.   Green corrosion inhibitors

2.1   Protection of iron based surfaces via green corrosion inhibitors

One of the most important uses of green inhibitors is the iron and steel industry. The inhibitors used in this sector are summarized below.

2.1.1   Protection of iron surfaces via green corrosion inhibitors

The first green corrosion inhibitors patented by Baldwin in 1895 are natural products such as *flour and yeast*. In the study conducted by Baldwin, it was observed that the corrosion effect was reduced in an acidic environment with a yield of 65-82% [32].

2.1.2   Protection of mild steel surfaces via green corrosion inhibitors

In 1985, Omar et al. investigated to usage of green inhibitors on mild steel. In the experiments carried out in acidic environment, the effects of *Papaia, Poinciana pulcherrima, Cassia occidentalis, Datura stramonium seeds, Calotropis procera, Azadirachta indica and Auforpio turkiale sap* materials were examined 94%, 96%, 94%, 93%, 98%, 84% and 69% yields were obtained respectively [33]. Zakvi and his colleagues examined the effect of *Swertia angustifolia* on mild steels in acidic conditions and determined that corrosion removal was achieved with a yield of 96% [34]. In the study carried out in acidic environment for the protection of the same material, *A. indica* was used as an inhibitor and a yield of 88% was obtained [35]. When the same inhibitor was used in NaCl medium, it was found that protection was achieved with 86% yield [36]. In the same study conducted by Quraishi and his team, the effects of *Punica granatum and Momordica charantia* inhibitors under the same conditions were examined and yields of 79% and 82% respectively [36].

In the studies on Mild steel, the effects of *Pongamia glabra and Annona squamosa* were investigated by Sakthivel et al. in acidic media and it was found that the material removed the corrosion with 95% yield [37]. In another study on corrosion removal in Mild steel, the effect of *Acacia arabica* was investigated by Verma in acidic media and it was found that corrosion was prevented with this inhibitor with 97% efficiency [38]. Also, *Allium cepa, Allium sativum, M. charantia* were used as an inhibitor in mild steel, and the material was found to provide 94% protection in acidic environment [39]. The effects of *Zanthoxylum alatum and Nypa fructicans* on corrosion protection of mild steel were investigated in acidic environment and it was observed that protection was achieved in 95% and 75% yields, respectively [40,41].

In studies on the protection of mild steel surfaces with green corrosion inhibitors in acidic media, E*xudate gum from Pachylobus edulis, Artemisia pallens and Corchorus olitorius* were used as natural inhibitors and yields of 56%, 98% and 93% respectively [42-44].

Finally, the effects of *Ricinus communis leaves* on NaCl media were investigated and it was determined that protection was provided with 84% yield [45].

The effect of green corrosion inhibitors on mild steel in acidic environment is given in Fig. 1.

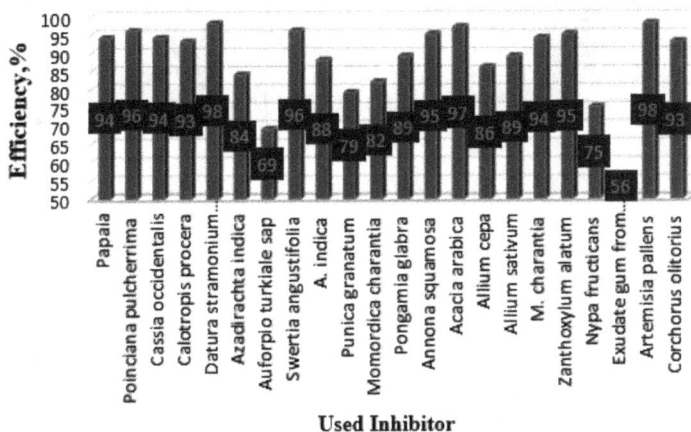

Figure 1. *Effect of green corrosion inhibitors on the mild steel in an acidic environment.*

The effect of green corrosion inhibitors on mild steel in NaCl medium is also shown in Fig. 2.

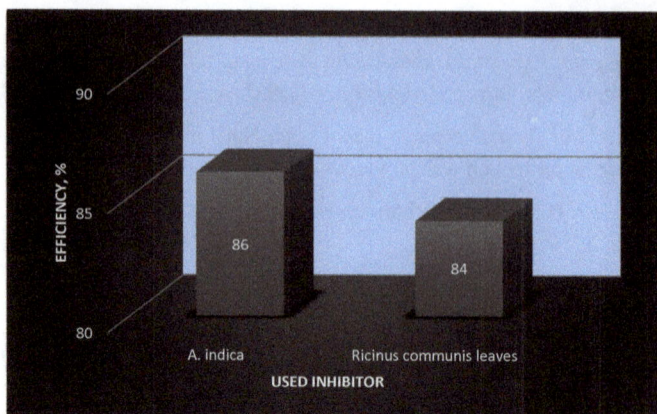

Figure 2.   *Effect of green corrosion inhibitors on the mild steel in NaCl environment.*

2.1.3 Protection of steel surfaces via green corrosion inhibitors

In the studies conducted for the protection of steel surfaces, the effects of herbs were first investigated in acidic media. *Thyme, coriander, hibiscus, anis, black cumin and cress* were used as inhibitors and yields ranging from 37% to 92% were obtained [46]. After examining the effect of herbs, corrosion behavior of steel materials in acidic media was investigated with *Eugenol from cloves, Acetyleugeno, Halfabar, Chamomile, Black cumin, Kidney bean* inhibitors, corrosion yields have been achieved 80%, 91%, 90%, 93%, 88% and 89% respectively [47,48]. Fig. 3 presents green corrosion inhibitors for steel surfaces in acidic media.

2.1.4 Protection of carbon steel surfaces via green corrosion inhibitors

As a green inhibitor of carbon steels, the effect of *Guar Gum* was studied by Abdallah in acidic media and protected by 94% yield. In another study on the application of carbon steel with natural inhibitors, the effect of *Caffeine and Nicotine* was investigated and 90% protection efficiency was obtained in the media containing chloride ions. For the same material, the effect of *Opuntia ficus indica* was studied in acidic environment and a protection of 91% was achieved [49-51]. Corrosion protection efficiencies obtained with green inhibitors used in carbon steels were examined and the findings are given in Fig. 4.

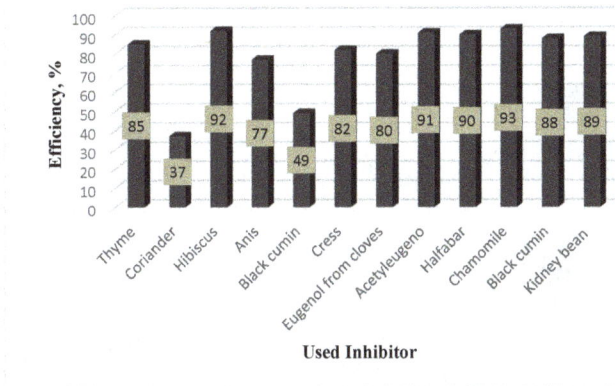

Figure 3.   Green corrosion inhibitors for steel surfaces in acidic media.

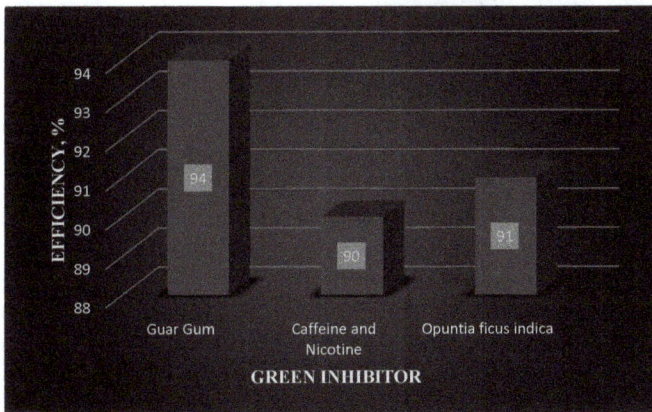

Figure 4.   Effect of green inhibitors on corrosion yield in carbon steels.

2.1.5   Protection of steel rebar surfaces via green corrosion inhibitors

Since steel rebar is used in the construction sector, research has been carried out in NaCl and alkali environment. In the studies carried out on steel rebar, *Bambusa Arundinacea, Leaves, Nuts, Bark, Tobacco, V. Amygdalina, Chamaerops Humilis L.* and *Morinda Lucida* were used as inhibitor obtained from natural sources, yields were found 85%,

72%, 91%, 70%, 78%, 91%, 42% and 93% respectively [52-55]. Inhibitors and yields used in steel rebar are presented in Fig. 5.

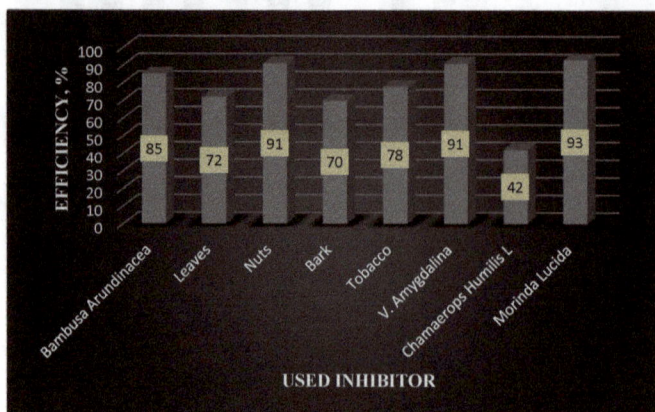

Figure 5.   *Inhibitors used in steel rebar and achieved corrosion protection efficiencies.*

2.2   Protection of aluminum surfaces via green corrosion inhibitors

Due to the light weight of aluminum (2.7 gr/cm^3 density), it finds many uses in the industry as an alternative to steel (7.8gr/cm^3density). However, the corrosion resistance of this material is not high like iron based surfaces. Environmental concerns are also important problems in aluminum protection against corrosion. The effects of green inhibitors on this subject, has become a popular topic among researchers. The first study on the protection of aluminum from corrosion by green inhibitors was carried out in 1972 by Saleh et al. In this study, *pomegranate peels* were carried out in acidic media and corrosion was prevented with 83% yield [56].

The natural inhibitors used for aluminum are *Opuntia Extract, Vernonia Amygdalina, Sansevieria Trifasciata, Exudate Gum, Gum Arabic* and *Raphia Hookeri*. Studies on this subject were carried out in HCl acid medium, 96%, 72%, 94%, 42%, 80% and 56% inhibition efficiencies obtained, respectively [57-61]. Sansevieria trifasciata was also studied in KOH environment and corrosion was prevented with 95% efficiency [59]. Corrosion inhibitation efficiencies of aluminum surfaces were given in Fig. 6.

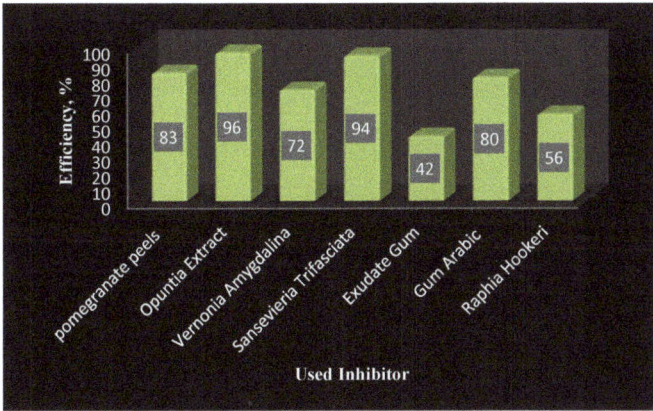

Figure 6.   Inhibitors used in aluminum and achieved corrosion protection efficiencies.

2.3   Protection of copper surfaces via green corrosion inhibitors

In the literature, the behavior of *Mangrove Tannin, Chitosan, Myrtus Communis, Tagetes Erecta, Alhagi Maurorum* and *Egyptian Licorice Extract* inhibitors in acidic media were investigated for copper surfaces. Accordingly, the inhibition yields were obtained 82%, 93%, 85%, 98%, 83% and 89% respectively [62-67]. These inhibitors efficiencies can found in Fig. 7.

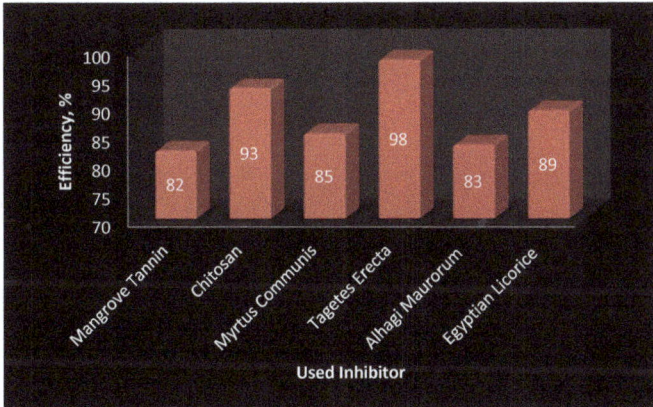

Figure 7.   Inhibitors used in copper and achieved corrosion protection efficiencies.

2.4   Protection of tin surfaces via green corrosion inhibitors

The first study on corrosion protection of tin with green inhibitors was carried out by Radojcic et al. In this study, *honey and radish juice* mixture was used as inhibitor and 93% inhibition efficiency was obtained [27]. In another study on the inhibitation of tin, the effect of *tomato peels* was examined and a yield of 73% was obtained in salt media [68].

2.5   Green corrosion inhibitors resources

The use of organic chemicals in anti-corrosion processes was first seen in the petroleum industry in the mid-20th century, and was also taught in the construction sector in the last decade of the same century. Yet, the toxic effects of the inhibitors used are an important problem. The use of natural inhibitors is an alternative method to eliminate the toxic and harmful effects of these materials. In this context, the use of green corrosion inhibitors is becoming popular day by day [69-72]. There are many green corrosion inhibitors, which concluded above, according to used media and surfaces. These inhibitors were obtained entirely from natural sources and classified in Fig. 8.

Figure 8.   Green Corrosion Inhibitors Sources.

3. Anti-corrosion mechanism (for natural inhibitors)

Corrosion is a process that takes shape according to standard free energy and progresses according to the ambient conditions and material structure. The material which has more negative Gibbs free energy than other, has a higher corrosion rate. When exposed to a corrosive environment, metals and alloys rust and stable corrosion products formed. Additional corrosion inhibitors are required to reduce the corrosion rate. The relative corrosion rate of any metal depends on the Pilling-Bedworth (PB) ratio defined as high temperature oxidation. According to this ratio, it is determined whether the film layer formed on the surface can prevent corrosion. Accordingly, if the volume of the metal is larger than the volume of the oxide layer formed by corrosion, the film layer contains various porosities. The opposite case, the protective film shows a holistic character, and metal, can protected from corrosion [73-75]. PB ratio is given in Equation 1.

$$M * d/n * m * D \qquad\qquad\qquad (1)$$

Where M and D are the molecular weight and density of corrosion product, m and d are the atomic weight and density of the metal, and n express the number of metallic atoms [28]. In inhibition applications firstly, the formed thin film or coating is expected to adsorb corrosive substances in the medium. There are many factors affecting the adsorption of the inhibitor on the metallic surface. As the inhibitor properties, mechanism of adsorption, chemical and electronic properties, temperature, electrolyte media and surface charge of the protected metals were determined as important factors affecting adsorption [76]. The adsorption mechanism can be physisorption, chemisorption or a combination of the both. The physical adsorption is the determined with $\Delta G°_{ads}$ (Gibbs free energy of aqueous solution). In the physorption process, $\Delta G°_{ads}$ should be maximum -20 kJ, if $\Delta G°_{ads}$ is more negative than this value, electron transfer occurs between the aggressive medium and the metal surface, which promotes corrosion [77].

The main mechanism of chemisorption is based on the inhibitor and the metal forming a covalent bond by electron transfer or sharing. It's $\Delta G°_{ads}$ value more negative than physisorption with maximum -40 kJ.mol^{-1} value [78]. By forming the aforementioned ionic and covalent bonds between the metal surface and the green inhibitor, the metallic surface is moved out of its active area and its corrosion is delayed. The passive film of the surface inhibitor is exposed to corrosive media and the substrate material is protected. Green inhibitors give good corrosion efficiency at room temperature and low temperatures, but their efficiency decreases in high temperature applications.

The effect of green inhibitors depends on the relationship between the organic coordination of plant species and the structure of the active ingredient. The anticorrosive activity resulting therefrom is based on the ability of these plant extracts compounds such as; alkaloids, flavonoids, etc. heterocyclic. Green Inhibitors and the available extracts are given in Table 1.

Table 1.   Green inhibitors and extracted products.

Green Inhibitor	Extracted Product
Garlic	Allyl Propyl Disulfide
Mustard Seeds	Alkaloid Berberine
Carrot	Pyrrolidine
Castor Seed	$C_8H_8N_2O_2$
$C_{10}H_{18}O$	Monomtrene-1,8-Cineole

3.1 Anodic, cathodic and mixed type inhibition

The inhibition effect may be anodic, cathodic or mixed. In the protection process performed with anodic inhibitors, the protective film formed disconnects the surface and the environment, while in the inhibition applications with cathodic protection, the precipitates formed perform the same function as the protective film. In Mixed-type inhibition processes, cathodic and anodic effects inhibit simultaneously by limiting electrochemical reactions. The inhibition mechanism is based on the inhibition of electron transfer by adsorbing the compounds containing N, P, S and O elements that are prone to electron exchange. Because these elements promote the transfer of electrons and this, encourage the corrosion. In order to provide these protections, the green inhibitors must be able to bind to the metal surface, which is achieved by adsorption. The interaction of organic materials with the surface is as follows:

- The interaction between the metal and the molecules that are not neutrally charged;

-Interaction of only double electrons of heteroatoms with empty d-orbital of metal;

Accordingly, the adsorption of inhibitors depends on the chemical and electrical properties of the metal surface. It is clear that for all the above-mentioned factors, the mechanism of inhibition needs to be carefully mentioned because corrosion of certain metals in certain environments is beneficial [79].

Theory and Applications of Green Corrosion Inhibitors   Materials Research Forum LLC
Materials Research Foundations **86** (2021) 38-62   https://doi.org/10.21741/9781644901052-2

4.   Corrosion inhibitors testing

Destructive and nondestructive inspection methods are used in green inhibitors as in other standard corrosion measurements to determine the amount of corrosion of metals. In order to fully characterize the corrosion inhibitors, some mechanical tests are required as well as standard methods [80].

The weight-loss method is a widely used method for calculating the corrosion rate of the material. In this method, the metallic surface is weighed and exposed to an aggressive surface. Afterwards, the corrosion products are cleaned from the surface and the part is re-weighed and the amount of corrosion is determined. Inhibitor efficiency is determined by comparing the corrosion rates in the absence and presence of the inhibitor [81]. The sample results examined by this method should contain an error bar, because defects caused by the material itself are capable of changing the results. Repeating experiments and evaluating the results are essential in weight loss applications.

Another method used to evaluate the efficiency of corrosion inhibitors is the electrochemical test technique. The method is very fast in comparison to the weight loss method in terms of measurement time. As a result of the evaluation made by electrochemical method, besides the corrosion resistance, data are obtained to shed light on corrosion prevention designs and applications. In the electrochemical method, a lot of data about anodic and cathodic behavior is obtained by making potentiodynamic analysis of metal and the media. Accordingly, it is determined which inhibitor can be used to protect metal and will be used in which environment. In order to obtain reliable results, the initial conditions of the measurement and the condition of the metal surface must be clearly defined during the sample preparation phase, and the tests shall be designed so that they do not vary from test to test. With the data obtained, the current-potential graph is plotted and the active, passive and passivation zones are determined [82]. Linear Polarization Resistance is one of the electrochemical measurement techniques. In this method, three electrodes designed as open circuit are polarized at +-10 mV and corrosion rate is determined according to circuit potential. This calculation is made by taking the potential slope against the current curve [83]. Electrochemical Impedance Spectroscopy (EIS) is used to calculate the amount of corrosion, the device performs its function by determining the polarization resistance. This device interprets the frequencies the sample gives against the AC voltage. This method is completed in minutes and is advantageous compared to similar methods that last for weeks. The data obtained in this method is used to create an equivalent circuit. The major disadvantage of the system is that it involves the creation of an equivalent circuit that models the metal solution interface, which is related to data interpretation [84].

The kinetic and thermodynamic data of adsorption can provide sufficient information to test the corrosion inhibition mechanism. Adsorption standard free energy (ΔG°_{ads}) is the most important parameter in determining the molecule's surface adsorption strenght. If the ΔG°_{ads} value is high, it is determined that the adsorption in the corrosive medium to the metal surface of the compound occurs strongly. ΔG°_{ads} can be calculated as $\Delta G^{\circ}_{ads}=$ ΔH°_{ads}(standard adsorption heat) -T(Temperature). ΔS°_{ads}(standard adsorption entropy). If the values found for enthalpy are negative, the adsorption of the inhibitory molecules is exothermic. The physical mechanism of this reaction is the active mechanism is the absorption. According to the Arrhenius equation, the activation energy of the adsorption can be calculated. If the required activation energy values are low, physical adsorption occurs [85].

5.   Economic and industrial opportunities

In order to understand whether the inhibition to be carried out to protect against corrosion is economical, the costs of damages caused by corrosion and the processes applied for protection must first be determined. While this issue has been on the agenda of countries since the 1950s, in 1978, cost items related to corrosion damages were reported from the US National Bureau of Standards. As a result of these concerns, corrosion management systems have been developed in industrial applications and costs have been controlled [86].

Costs caused by corrosion are listed as the reconstruction of buildings, material losses during production, maintenance costs, insurance costs, machine backup obligation, use of inhibitors, research and development processes, training, human resources and salaries of these people. According to the report published by the National Association of Corrosion Engineers in 2013, the cost of corrosion worldwide covers 3.4% of the gross national product. This value is calculated as approximately 2.5 trillion US $. In the same report, 35% savings can be achieved by developing corrosion management systems, was reported. This corresponds to approximately US$ 875 billion [87]. These savings amounts are encouraging for corrosion inhibition applications.

A lot of research has been done on corrosion applications, and various patents have been obtained on the subject of green inhibitor. The patents obtained in the last ten years are given in Table 2.

Table 2. Green inhibitors patents with economic aspects.

Green Inhibitor	Protected surface	Economic aspect	Ref.
Phoenix Clactylifera	Steel	Bio Waste	[88]
Tobacco,	Oil Transportation Pipelines	Microbiologically Corrosion	[89]
Sweet Potato Stems, Lettuce Flower Stalks	Metals	Bio Waste	[90]
Fruit Extracts (Mango, Cashew, Passion-Fruit and Orange)	Steel	Bio Waste	[91]
Plant Extracts	Aluminum, Aluminum Alloys, Copper and Copper Alloys	Bio Waste	[92]
Vegetable Oil	petrochemical industry	Bio Waste	[93]

Considering the savings made to avoid corrosion, researchers' inclination towards reasonable products such as bio-waste, plant extracts as an inhibitor will provide significant added value to this area. Increases in the gains mentioned in the report of the US National Bureau of Standard will be possible with the studies in this direction.

References

[1] M. Drahansky, M. Paridah, A. Moradbak, A. Mohamed, F. Abdulwahab, T. Owolabi, M. Asniza, S.H. Abdul Khalid, We are IntechOpen, the world's leading publisher of open access books built by scientists, for scientists TOP 1 %, Intech. i (2016) 13.

[2] N.T. Nassar, T.E. Graedel, E.M. Harper, By-product metals are technologically essential but have problematic supply, Sci. Adv. 1 (2015) 1–10. https://doi.org/10.1126/sciadv.1400180

[3] P.A. Schweitzer, Encyclopedia of Corrosion Technology, Marcel Dekker Inc., New York, 2004, pp. 1-73.

[4] N. Eliaz, Corrosion of metallic biomaterials: A review, Materials (Basel). 12(3) (2019), 407. https://doi.org/10.3390/ma12030407

[5] I. Gurrappa, G. Malakondaiah, Effect of environment on corrosion characteristics of newly developed DMR-1700 structural steel, Sci. Technol. Adv. Mater. 9(2) (2008) 025005-025012. https://doi.org/10.1088/1468-6996/9/2/025005

[6]   M.J. Pryor, L.L. Shreir, Bimetallic corrosion, Corros. 1 (1976) 192-214. https://doi.org/10.1016/B978-0-408-00109-0.50016-X

[7]   J.L. Smith, Y.P. Virmani, Materials and Methods for Corrosion Control of Reinforced and Prestressed Concrete Structures in New Construction, Fed. Highw. Adm. U.S. Dep. Transp. (2000) 82.

[8]   M. Abdallah, B.H. Asghar, I. Zaafarany, A.S. Fouda, The inhibition of carbon steel corrosion in hydrochloric acid solution using some phenolic compounds, Int. J. Electrochem. Sci. 7 (2012) 282–304.

[9]   Y. Fujino, Corrosion protection of steel bridges, Corros. Eng. 38 (1989) 595–606. https://doi.org/10.3323/jcorr1974.38.10_546

[10]   G.T. Hefter, N.A. North, S.H. Tan, Organic corrosion inhibitors in neutral solutions; part 1 - inhibition of steel, copper, and aluminum by straight chain carboxylates, Corros. 53 (1997) 657–667. https://doi.org/10.5006/1.3290298

[11]   A.A.F. Sabirneeza, R. Geethanjali, S. Subhashini, Polymeric corrosion inhibitors for iron and its alloys: A review, Chem. Eng. Commun. 202 (2015) 232–244. https://doi.org/10.1080/00986445.2014.934448

[12]   F. Branzoi, V. Branzoi, C. Licu, Corrosion inhibition of carbon steel in cooling water systems by new organic polymers as green inhibitors, Mater. Corros. 65 (2014) 637–647. https://doi.org/10.1002/maco.201206579

[13]   R.A. Sheldon, Metrics of green chemistry and sustainability: past, present, and future, ACS Sustain. Chem. Eng. 6 (2018) 32–48. https://doi.org/10.1021/acssuschemeng.7b03505

[14]   P.T. Anastas, M.M. Kirchhoff, Origins, current status, and future challenges of green chemistry, Acc. Chem. Res. 35 (2002) 686–694. https://doi.org/10.1021/ar010065m

[15]   F. Hollmann, J. Tao, R. J. Kazlauskas: Biocatalysis for green chemistry and chemical process development, Catal. Letters. 143 (2013) 982–982. https://doi.org/10.1007/s10562-013-1036-0

[16]   F. Chemat, M. Vian, Alternative Solvents for Natural Products Extraction, Green Chemistry and Sustainable Technology, Springer, 2014. https://doi.org/10.1007/978-3-662-43628-8

[17] R.A. Sheldon, Green chemistry, catalysis and valorization of waste biomass, J. Mol. Catal. A Chem. 422 (2016) 3–12. https://doi.org/10.1016/j.molcata.2016.01.013

[18] R. Hartman, R. Helmy, M. Al-Sayah, C.J. Welch, Analytical Method Volume Intensity (AMVI): A green chemistry metric for HPLC methodology in the pharmaceutical industry, Green Chem. 13 (2011) 934–939. https://doi.org/10.1039/c0gc00524j

[19] C.J. Li, B.M. Trost, Green chemistry for chemical synthesis, Proc. Natl. Acad. Sci. U. S. A. 105 (2008) 13197–13202. https://doi.org/10.1073/pnas.0804348105

[20] R. Höfer, J. Bigorra, Biomass-based green chemistry: Sustainable solutions for modern economies, Green Chem. Lett. Rev. 1 (2008) 79–97. https://doi.org/10.1080/17518250802342519

[21] M. Ismail, A.S. Abdulrahman, M.S. Hussain, Solid waste as environmental benign corrosion inhibitors in acid medium, Int. J. Eng. Sci. Technol. 3 (2011) 1742–1748.

[22] N.A. Odewunmi, S.A. Umoren, Z.M. Gasem, Watermelon waste products as green corrosion inhibitors for mild steel in HCl solution, J. Environ. Chem. Eng. 3 (2015) 286–296. https://doi.org/10.1016/j.jece.2014.10.014

[23] S. Antonio, Corrosion inhibition efficiencies of heterocyclic, unsaturated (aromatic and nonaromatic) compounds (pyrimidines, matic part, pyridines, and quinolines) were correlated with quantum chemical indices of the respective molecules . Inhibition Effic, 6 (n.d.).

[24] S. Marzorati, L. Verotta, S.P. Trasatti, Green corrosion inhibitors from natural sources and biomass wastes, Molecules. 24(1) (2019) 48 PMC6337343. https://doi.org/10.3390/molecules24010048

[25] G. Bereket, A. Yurt, The inhibition effect of amino acids and hydroxy carboxylic acids on pitting corrosion of aluminum alloy 7075, Corros. Sci. 43 (2001) 1179–1195. https://doi.org/10.1016/S0010-938X(00)00135-9

[26] B.E.A. Rani, B.B.J. Basu, Green inhibitors for corrosion protection of metals and alloys: An overview, Int. J. Corros. 2012 (2012) 380217. https://doi.org/10.1155/2012/380217

[27] I. Radojčić, K. Berković, S. Kovač, J. Vorkapić-Furač, Natural honey and black radish juice as tin corrosion inhibitors, Corros. Sci. 50 (2008) 1498–1504.

https://doi.org/10.1016/j.corsci.2008.01.013

[28] K. Krishnaveni, J. Ravichandran, Effect of aqueous extract of leaves of Morinda tinctoria on corrosion inhibition of aluminium surface in HCl medium, Oral Oncol. 50 (2014) 2704–2712. https://doi.org/10.1016/S1003-6326(14)63401-4

[29] P. Singh, V. Srivastava, M.A. Quraishi, Novel quinoline derivatives as green corrosion inhibitors for mild steel in acidic medium: Electrochemical, SEM, AFM, and XPS studies, J. Mol. Liq. 216 (2016) 164–173. https://doi.org/10.1016/j.molliq.2015.12.086

[30] B.S. Sanatkumar, J. Nayak, A. Nityananda Shetty, Influence of 2-(4-chlorophenyl)-2-oxoethyl benzoate on the hydrogen evolution and corrosion inhibition of 18 Ni 250 grade weld aged maraging steel in 1.0 M sulfuric acid medium, Int. J. Hydrogen Energy. 37 (2012) 9431–9442. https://doi.org/10.1016/j.ijhydene.2012.02.165

[31] K.M. Manamela, L.C. Murulana, M.M. Kabanda, E.E. Ebenso, Adsorptive and DFT studies of some imidazolium based ionic liquids as corrosion inhibitors for zinc in acidic medium, Int. J. Electrochem. Sci. 9 (2014) 3029–3046.

[32] P.C. Okafor, M.E. Ikpi, I.E. Uwah, E.E. Ebenso, U.J. Ekpe, S.A. Umoren, Inhibitory action of Phyllanthus amarus extracts on the corrosion of mild steel in acidic media, Corros. Sci. 50 (2008) 2310–2317. https://doi.org/10.1016/j.corsci.2008.05.009

[33] I.H. Omar, Plant extracts as corrosion inhibitors of mild steel in HCl solutions, Surf. Tech. 24 (1985) 391–399. https://doi.org/10.1016/0376-4583(85)90057-3

[34] S.J. Zakvi, G.N. Mehta, Acid corrosion of mild steel and its inhibition by Swertia angustifolia study by electrochemical techniques, Transactions of the SAEST. 23(4) (1988) 407-410.

[35] U.J. Ekpe, E.E. Ebenso, U.J. Ibok, Inhibitory action of Azadirachta indica leaves extract on the corrosion of mild steel in H_2SO_4. J. West Afr. J. Biol. Appl. Chem. 37 (1994) 13-30.

[36] M.A. Quraishi, I.H. Farooqi, P.A.Saini, Investigation of some green compounds as corrosion and scale inhibitors for cooling systems. Corrosion. 55(5) (1999) 493-497. https://doi.org/10.5006/1.3284011

[37] P. Sakthivel, P.V. Nirmala, S. Umamaheswari, A.A.A. Antony, G. Paruthimal

Kalaignan, A. Gopalan, T. Vasudevan, Corrosion inhibition of mild steel by extracts of Pongamia glabra and Annona squamosa in acidic media, Bull. Electrochem. 15(2) (1999) 83-86.

[38] S.A. Verma, G.N. Mehta, Effect of acid extracts of Acacia arabica on acid corrosion of mild steel. Bull. Electrochem. 15(2) (1999) 67-70.

[39] K.S. Parikh, K.J. Joshi, Natural compounds onion (Allium cepa), garlic (Allium sativum) and bitter gourd (Momordica charantia) as corrosion inhibitors for mild steel in hydrochloric acid. Trans. Soc. Adv. Electrochem. Sci. Technol. 39(1/2) (2004) 29-35.

[40] G. Gunasekaran, L.R. Chauhan, Eco friendly inhibitor for corrosion inhibition of mild steel in phosphoric acid medium. Electrochim. Acta. 49(25) (2004) 4387-4395. https://doi.org/10.1016/j.electacta.2004.04.030

[41] K.O. Orubite, N.C. Oforka, Inhibition of the corrosion of mild steel in hydrochloric acid solutions by the extracts of leaves of Nypa fruticans Wurmb. Mater. Lett. 58(11) (2004) 1768-1772. https://doi.org/10.1016/j.matlet.2003.11.030

[42] S.A. Umoren, U.F. Ekanem, Inhibition of mild steel corrosion in H_2SO_4 using exudate gum from Pachylobus edulis and synergistic potassium halide additives, J. Chem. Eng. Comm. 197(10) (2010) 1339-1356. https://doi.org/10.1080/00986441003626086

[43] S. Garai, S. Garai, P. Jaisankar, J.K. Singh, A. Elango, A comprehensive study on crude methanolic extract of Artemisia pallens (Asteraceae) and its active component as effective corrosion inhibitors of mild steel in acid solution, Corros. Sci. 60 (2012) 193-204. https://doi.org/10.1016/j.corsci.2012.03.036

[44] M. Gobara, B. Zaghloul, A. Baraka, M. Elsayed, M. Zorainy, M.M. Kotb, H. Elnabarawy, Green corrosion inhibition of mild steel to aqueous sulfuric acid by the extract of Corchorus olitorius stems, Mater. Res. Express. 4(4) (2017) 046504. https://doi.org/10.1088/2053-1591/aa664a

[45] R. Sathiyanathan, S. Maruthamuthu, M. Selvanayagam, S. Mohanan, N.Palaniswamy, Corrosion inhibition of mild steel by ethanolic extracts of Ricinus communis leaves, Indian J. Chem. Technol. 12 (2005) 356-360.

[46] E. Khamis, N. Alandis, Herbs as new type of green inhibitors for acidic corrosion of steel. Material Wissenschaft und Werkstofftechnik, 33(9) (2002) 550-554. https://doi.org/10.1002/1521-4052(200209)33:9<550::AID-MAWE550>3.0.CO;2-

G

[47] E. Chaieb, A. Bouyanzer, B. Hammouti, M. Benkaddour, Inhibition of the corrosion of steel in 1 M HCl by eugenol derivatives, Appl. Surf. Sci. 246(1) (2005) 199-206. https://doi.org/10.1016/j.apsusc.2004.11.011

[48] A.M. Abdel-Gaber, B.A. Abd-El-Nabey, I.M. Sidahmed, A.M. El-Zayady, M. Saadawy, Inhibitive action of some plant extracts on the corrosion of steel in acidic media, Corros. Sci. 48(9) (2006) 27652779. https://doi.org/10.1016/j.corsci.2005.09.017

[49] M. Abdallah, Guar gum as corrosion inhibitor for carbon steel in sulfuric acid solutions. Port. Electrochim. Acta. 22(2) (2004) 161-175. https://doi.org/10.4152/pea.200402161

[50] J.P. Flores-De los Ríos, M. Sánchez-Carrillo, C.G. Nava-Dino, J.G. Chacón-Nava, G. Rodríguez, E. Huape-Padilla, M.A. Neri-Flores, A. Martínez-Villafañe, Opuntia ficusindica extract as green corrosion inhibitor for carbon steel in 1 M HCl solution, J. Spectros. 2015 (2015) 14692-14701. https://doi.org/10.1155/2015/714692

[51] E. Vuorinen, E. Kálmán , W. Focke, Introduction to vapour phase corrosion inhibitors in metal packaging, Surf. Eng. 20(4) (2004) 281-284. https://doi.org/10.1179/026708404225016481

[52] C.A. Loto, R.T. Loto, A.P.I. Popoola, Electrode potential monitoring of effect of plants extracts addition on the electrochemical corrosion behaviour of mild steel reinforcement in concrete, Int. J. Electrochem. Sci. 6 (2011) 3452-3465.

[53] C.A. Loto, O.O. Joseph, R.T. Loto, A.P.I. Popoola, Inhibition effect of Vernonia amygdalina extract on the corrosion of mild steel reinforcement in concrete in 3.5M NaCl environment, Int. J. Electrochem. Sci. 8 (2013) 11087-11100.

[54] D.B. Left, M. Zertoubi, S. Khoudali, M. Benaissa, A. Irhzo, M. Azzi, Effect of methanol extract of Chamaerops humilis L. leaves (MECHLL) on the protection performance of oxide film formed on reinforcement steel surface in concrete simulated pore solution, Int. J. Electrochem. Sci. 8(10) (2013) 11768-11781.

[55] J.O. Okeniyi, C.A. Loto, A.P.I. Popoola., Morinda lucida effects on steel-reinforced concrete in 3.5% NaCl: implications for corrosion-protection of wind-energy structures in saline/marine environments, Energ. Proced. 50 (2014) 421-428. https://doi.org/10.1016/j.egypro.2014.06.051

[56] R.M. Saleh, A.A. El-Hosary, Corrosion inhibition by naturally occurring substances. Effect of Pomegranate Juice and the Aqueous Extracts of Pomegranate Fruit Shells, Tamarind Fruits and Tea Leaves on the Corrosion of Al, in: El-Hosary (Editor), Electrochemistry; thirteenth ed., Karaikudi, CECRI, 1972.

[57] A.Y. El-Etre Inhibition of aluminum corrosion using Opuntia extract, Corros. Sci. 45(11) (2003) 2485-2495. https://doi.org/10.1016/S0010-938X(03)00066-0

[58] G.O. Avwiri, F.O. Igho, Inhibitive action of Vernonia amygdalina on the corrosion of aluminium alloys in acidic media, Mater. Lett. 57(22) (2003) 3705-3711. https://doi.org/10.1016/S0167-577X(03)00167-8

[59] E.E.Oguzie, Corrosion inhibition of aluminium in acidic and alkaline media by Sansevieria trifasciata extract, Corros. Sci. 49(3) (2007) 1527-1539. https://doi.org/10.1016/j.corsci.2006.08.009

[60] S.A. Umoren, I.B. Obot, E.E. Ebenso, N. Obi-Egbedi, Studies on the inhibitive effect of exudate gum from Dacroydes edulis on the acid corrosion of aluminium, Port. Electrochim. Acta. 26(2) (2008) 199-209. https://doi.org/10.4152/pea.200802199

[61] S.A. Umoren, I.B. Obot, E.E. Ebenso, N. Obi-Egbedi, The inhibition of aluminium corrosion in hydrochloric acid solution by exudate gum from Raphia hookeri, Desalination. 247(1-3) (2009) 561-572. https://doi.org/10.1016/j.desal.2008.09.005

[62] A.M. Shah, A.A. Rahim, S.A. Hamid, S. Yahya, Green inhibitors for copper corrosion by mangrove tannin. Int. J. Electrochem. Sci. 8 (2013) 2140-2153.

[63] M.N.El-Haddad, Chitosan as a green inhibitor for copper corrosion in acidic medium. Int. J. Biol. Macromolecules. 55 (2013) 142-149. https://doi.org/10.1016/j.ijbiomac.2012.12.044

[64] M. Bozorg, T. Shahrabi Farahani, J. Neshati, Z. Chaghazardi, G. Mohammadi Ziarani, Myrtus communis as green inhibitor of copper corrosion in sulfuric acid, Ind. Eng. Chem. Res. 53(11) (2014) 4295-4303. https://doi.org/10.1021/ie404056w

[65] P. Mourya, S. Banerjee, M.M. Singh, Corrosion inhibition of mild steel in acidic solution by Tagetes erecta (marigold flower) extract as a green inhibitor, Corros. Sci. 85 (2014) 352-363. https://doi.org/10.1016/j.corsci.2014.04.036

[66] B.A. Abd-El-Nabey, S. El-Housseiny, G.A. El-Naggar, E.A. Matter, G. Esmail, Inhibitive action of Alhagi maurorum plant extract on the corrosion of copper in

Materials Research Forum LLC
https://doi.org/10.21741/9781644901052-2

0.5 M H_2SO_4 Phys. Chem. 5(3) (2015) 49-62.

[67] M.A. Deyab, Egyptian licorice extract as a green corrosion inhibitor for copper in hydrochloric acid solution. J. Ind. Eng. Chem. 22 (2015) 384-389. https://doi.org/10.1016/j.jiec.2014.07.036

[68] A.N. Grassino, J. Halambek, S. Djakovi´c, S.R. Brn˘ci´c, M. Dent, Z. Grabari´c, Utilization of tomato peel waste from canning factory as a potential source for pectin production and application as tin corrosion inhibitor, Food Hydrocoll. 52 (2016) 265–274. https://doi.org/10.1016/j.foodhyd.2015.06.020

[69] R. Rosliza, W.B.W. Nik, Improvement of corrosion resistance of AA6061 alloy by tapioca starch in seawater, Curr. Appl. Phys. 10 (2010) 221–229. https://doi.org/10.1016/j.cap.2009.05.027

[70] E.E. Oguzie, Evaluation of the inhibitive effect of some plant extracts on the acid corrosion of mild steel, Corros. Sci. 50 (2008) 2993–2998. https://doi.org/10.1016/j.corsci.2008.08.004

[71] A. Ostovari, S.M. Hoseinieh, M. Peikari, S.R. Shadizadeh, S.J. Hashemi, Corrosion inhibition of mild steelin 1 M HCl solution by henna extract: A comparative study of the inhibition by henna and its constituents(Lawsone, Gallic acid,α-d-Glucose and Tannic acid), Corros. Sci. 51 (2009) 1935–1949. https://doi.org/10.1016/j.corsci.2009.05.024

[72] E.E. Oguzie, K.L. Oguzie, C.O. Akalezi, I.O. Udeze, J.N. Ogbulie, V.O. Njoku. Natural products for materials protection: Corrosion and microbial growth inhibition using Capsicum frutescens biomass extracts, ACS Sustain. Chem. Eng. 1 (2013) 214–225. https://doi.org/10.1021/sc300145k

[73] R.E. Bedworth, N.B.Pilling, The oxidation of metals at high temperatures, J. Inst. Met. 29(3) (1923) 529-582.

[74] S. Manimegalai, P. Manjula, Thermodynamic and adsorption studies for corrosion inhibition of mild steel in aqueous media by Sargasam swartzii (brown algae), J. Mater. Environ. Sci. 6(6) (2015) 1629-1637.

[75] Y. Roh, S.Y. Lee, M.P.Elless, Characterization of corrosion products in the permeable reactive barriers, Environ. Geology. 40(1-2) (2000) 184-194. https://doi.org/10.1007/s002540000178

[76] A.K. Maayta, N.A.F. Al-Rawashdeh, Inhibition of acidic corrosion of pure

aluminum by some organic compounds, Corros. Sci. 46(5) (2004) 1129-1140. https://doi.org/10.1016/j.corsci.2003.09.009

[77] B.B. Damaskin, O.A. Petrii, V.V. Batrakov, Adsorption, New York: Plenum Press, 1971.

[78] S. Martinez, I. Štern, Thermodynamic characterization of metal dissolution and inhibitor adsorption processes in the low carbon steel/mimosa tannin/sulfuric acid, Appl. Surf. Sci. 199(1) (2002) 83-89. https://doi.org/10.1016/S0169-4332(02)00546-9

[79] M.V. Fiori-Bimbi, P.E. Alvarez, H. Vaca, C.A. Gervasi, Corrosion inhibition of mild steel in HCL solution by pectin, Corros. Sci. 92 (2015) 192–199. https://doi.org/10.1016/j.corsci.2014.12.002

[80] A.Y. El-Etre, M. Abdallah, Z.E. El-Tantawy, Corrosion inhibition of some metals using lawsonia extract, Corros. Sci. 47 (2005) 385–395. https://doi.org/10.1016/j.corsci.2004.06.006

[81] S. Papavinasam, Evaluation and selection of corrosion inhibitors, Uhlig's Corros. Handb. (2000) 1169–1178.

[82] M. Faustin, A. Maciuk, P. Salvin, C. Roos, M. Lebrini, Corrosion inhibition of C38 steel by alkaloids extract of Geissospermum laeve in 1M hydrochloric acid: Electrochemical and phytochemical studies, Corros. Sci. 92 (2015) 287–300. https://doi.org/10.1016/j.corsci.2014.12.005

[83] M.A. Stern, Method for determining corrosion rates from linear polarization data, Corros. 14 (1958) 60–64. https://doi.org/10.5006/0010-9312-14.9.60

[84] L.M.P. Dolabella, J.G. Oliveira, V. Lins, T. Matencio, W.L. Vasconcelos, Ethanol extract of propolis as a protective coating for mild steel in chloride media, J. Coat. Technol. Res. 13 (2016) 543–555. https://doi.org/10.1007/s11998-015-9765-1

[85] A. Fawzy, M. Abdallah, I.A. Zaafarany, S.A. Ahmed, I.I. Althagafi, Thermodynamic, kinetic and mechanistic approach to the corrosion inhibition of carbon steel by new synthesized amino acids-based surfactants as green inhibitors in neutral and alkaline aqueous media, J. Mol. Liq. 265 (2018) 276–291. https://doi.org/10.1016/j.molliq.2018.05.140

[86] G. Bianchi, F. Mazza. Corrosione e Protezione dei Metalli; Associazione Italiana di Metallurgia, Milano, Italy, 2000.

[87] G.Koch, J. Varney, N. Thompson, O. Moghissi, M. Gould, J. Payer, International Measures of Prevention, Application, and Economics of Corrosion Technologies Study, NACE Internationa, Houston, TX, USA, 2016.

[88] G. Rexin Thusnavis, K.P. Vinod Kumar, Green Corrosion Inhibitor for Steel in Acid Medium, 2014. Application No. 6278/CHE/2014 A.

[89] Indian Oil Corporation Limited, Naturally derived Corrosion Inhibitors Composition, Process for Preparing the same and use Thereof 2008. Patent Application 10 March 2008.

[90] Extract Corrosion Inhibitor of Sweet Potato Stems and Lettuce Flower Stalks and Preparation Method thereof. Patent No. CN102492948B, 31 July 2013.

[91] J.A. Ponciano Gomes, J. Cardoso Rocha, E. D'Elia, Use of Fruit Skin Extracts as Corrosion Inhibitors and Process for Producing Same, U.S. Patent US8926867B2, 6 January 2015.

[92] P.J. Kinlen, L.Pinheiro Santos Pimenta, Methods and Apparatuses for Selecting Natural Product Corrosion Inhibitors for Application to Substrates, U.S. Patent Application No. US 2018/0202051 Al, 19 July 2018.

[93] R.Lima, A. Casalini, A. Palumbo, G.Rivici, Corrosion Inhibitor Comprising Complex Oligomeric Structures derived from Vegetable Oils, Patent Application No. WO 2017/140836 Al, 24 August 2017.

Materials Research Forum LLC
https://doi.org/10.21741/9781644901052-3

Chapter 3

Green Inhibitors for Biocorrosion and Prevention

H.M. Saleh*, A.I. Hassan

Radioisotope Department, Nuclear Research Center, Atomic Energy Authority, Dokki, 12311, Giza, Egypt

*hosam.saleh@eaea.org.eg, hosamsaleh70@yahoo.com

Abstract

Material corrosion can be a limiting factor for various elements in many applications. Accordingly, corrosion operations must be avoided and restricted or diminish the associated damage. The corrosion rate is one of the most important characteristics of corrosion processes, that are difficult to measure in the non-aqueous environment. Indeed, the use of corrosion inhibitors is a known counterstrategy when it needs to prevent, control, or delay mineral corrosion. In this regard, green chemistry emphasizes the substantivized of maintaining the environment and human health safely and economically manner at the same time, which aims to avoid toxins and reduce waste. The field of mineral degradation, which is generally faced with the use of toxic compounds, has found fertile research in green chemistry. The field of mineral degradation, which generally faces the use of toxic compounds, has found fertile research in green chemistry. It is worth noting that green inhibitors are the best as they are environmentally biodegradable, and renewable. In this chapter, we will demonstrate the causes of corrosion and the required processes to restrain them, and the competence of corrosion inhibitors, which can promote reducing the damage provoked by corrosion. Besides, we will highlight the importance of unique innovations that act as corrosion inhibitors being an environmentally friendly natural inhibitor.

Keywords

Biocorrosion, Green Inhibitors, Environmentally Friendly, Prevention

Contents

Theory and Applications of Green Corrosion Inhibitors
Materials Research Foundations **86** (2021) 63-90

Materials Research Forum LLC
https://doi.org/10.21741/9781644901052-3

1. Introduction

Corrosion is the breakdown of substances due to interaction with the surrounding environment. There are other definitions of corrosion such as the breakdown of materials by a non-mechanical act, as well as the reverse extract of metals [1]. Corrosion is a science of technology, concerned with the study of matter and its outer environment, and the interrelation between them of physical and chemical effects, which cause changes to metal, and lead to its damage, or change its specifications in terms of mechanical and chemical resistance [1].

The environment is the main factor in the corrosion of metals. In practically all environments and media, corrosion occurs to varying degrees. Examples include moist air, clean water, salt-containing, rural and urban climate, industrial zone air, gas vapor such as chlorine, ammonia, hydrogen sulfide, and sulphur dioxide, fuel gases such as carbon monoxide and carbon dioxide, and mineral acids such as hydrochloric, sulfuric, nitric, and acids organic materials such as acetic and formic acid, and organic solvents, oils, and many types of food [2]. It is noted that inorganic substances cause more metal breakdown than organic materials. For example, corrosion in industrial petroleum complexes is caused by more sodium chloride, sulphur, hydrochloric acid, sulfuric and water than from petroleum derivatives [3].

Corrosion Engineering is the process of applying science and practical skills to prevent or control mineral breakdowns safely and economically [3]. For the corrosion engineer to perform his work, he must be familiar with the basics of corrosion and its applications, which include several sciences, including the basics of chemistry, especially electrochemistry, and science [4].

Minerals, material properties, as well as corrosion tests, the nature of the environment, availability and manufacture of materials, and design [5].

The importance of studying the impact of corrosion on industry falls under the basic areas of cost reduction and includes reducing the use or loss of materials due to corrosion such as pipelines, tanks, machinery, ships, bridges, sidewalks, and all-metal structures. Improving operational safety, by preserving lives and property from sudden breakdowns such as boilers, gas tanks, radioactive material-containing structures, aircraft, and moving parts of the machines [5]. Preserving mineral resources, as these resources are limited in the world, and their waste includes wasting energy and water in addition to the material itself and its manufacture [5].

1.1 The portability of the metal to the corrosion

The difference in minerals in terms of their ability to corrosion is due to two factors.

Theory and Applications of Green Corrosion Inhibitors   Materials Research Forum LLC
Materials Research Foundations **86** (2021) 63-90   https://doi.org/10.21741/9781644901052-3

The basic two are:

The chemical activity of the metal:

The chemical activity of the metal, as the corrosion ability of the metal increases with increasing its chemical activity, and the smaller its activity, the greater its resistance to corrosion. Returning to the electrochemical chain.

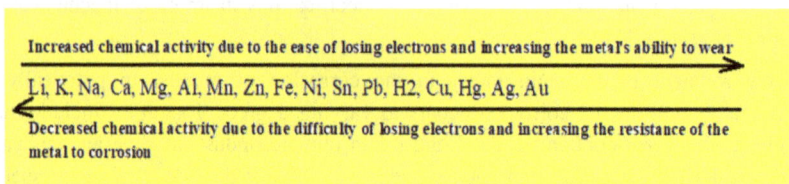

Increased chemical activity due to the ease of losing electrons and increasing the metal's ability to wear

→

Li, K, Na, Ca, Mg, Al, Mn, Zn, Fe, Ni, Sn, Pb, H2, Cu, Hg, Ag, Au

←

Decreased chemical activity due to the difficulty of losing electrons and increasing the resistance of the metal to corrosion

Figure 1. The corrosion ability of metals.

Table 1.   *Oxidation potential of some metals in the electrochemical chain.*

Metal	Ions	Oxidative stress (volt)
K	K^+	2.92
Ca	Ca^{++}	2.84
Na	Na^+	2.713
Mg	Mg^{++}	2.36
Al	Al^{+++}	1.66
Mn	Mn^{++}	1.05
Zn	Zn^{++}	0.763
Fe	Fe^{++}	0.44
Ni	Ni^{++}	0.23
Sn	Sn^{++}	0.14
Pb	Pb^{++}	0.126
H2	$H2^+$	0.000
Cu	Cu^{++}	0.34-
Hg	Hg^{++}	0.798-
Ag	Ag^+	0.799-
Au	Au^{+++}	1.42-

Potential voltage:

Potential voltage is very beneficial because they help to know the metal's exposure to corrosion by a certain solution. If the oxidation voltage is negative, this means that the metal will not oxidize with acids, and if the voltage is positive, the metal will oxidize with acids [2,7]. In Table 1, we note that iron has a positive voltage oxidized by acids releasing hydrogen, while gold, whose voltage is negative, resists corrosion with acids.

1.2   The factors affecting the speed of corrosion

There are two types, the first concerning the substance that suffers from corrosion, the strength bonding between its atoms or its particles, the shape and pattern of its crystalline structure, the defects and the impurities which it contains. It also includes the mechanical or thermal effects or their defects to which the corrosive substance has been exposed during or after the preparation and manufacturing process [8].

The second type includes factors that affect the velocity of corrosion, that is, the environment that causes this corrosion. It includes the nature of the corrosive substances, the concentration, the effect of the pH to which the surface of the material is exposed, and the presence or absence of oxygen in the reaction medium [9] This type of factor extends to include temperature, pressure and the presence or absence of accelerated or obstructive materials in the environment of corrosion [10]. If the corrosion reaction occurs between a metal and a metal, for example, so that the metal is oxidized and the metal is reduced, and the reaction is formed in the absence of a liquid solvent, the corrosion is of a dry type. As for nonmetals in such corrosion, they usually represent oxygen, halogen, sulfur vapor, hydrogen sulfide gas, hydrogen chloride gas, or other effective gases or fumes, and the oxide, halide, or sulfide layer is a corrosive process. The corrosion reaction may be of a wet type, as it occurs in the presence of a liquid such as in a solution, and the oxidation (or reaction) of the metal (or alloy) occurs at certain surface areas. The corrosion process permeates the transfer of electrons from the elevator (anode) that represents the metal (or alloy) oxidized to the cathode (cathode), which represents the electrochemical reduction sites formed in the process [11,12].

1.3   Types of corrosions

1.3.1 Pure chemical corrosion

This corrosion occurs because of a normal chemical reaction between the medium and the surface, which allows the radicals are exchanged [13]. So that the corrosion product is a new substance that dissolves in the outer medium, and it may be in the form of a volatile gaseous substance or a substance that does not dissolve in the corrosive medium or fragile material that does not stick well to the surface of the metal to allow the corrosion to reach new layers of the metal surface [14]. For example, the effect of acids on concrete and converting it into dissolved material:

$$CaCO_3 + 2HCL \longrightarrow CaCl_2 + CO_2 + H_2O \qquad (1)$$

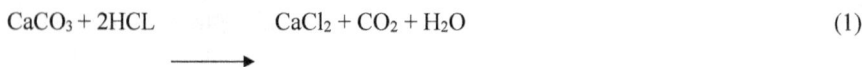

The effect of acid on the metals:

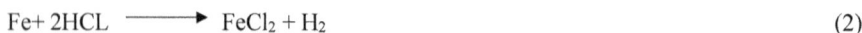

$$Fe + 2HCL \longrightarrow FeCl_2 + H_2 \qquad (2)$$

Calcium chloride and ferrous chloride are soluble substances in water; thus, iron and concrete are corrosive [15]. A common and serious chemical corrosion is the effect of carbon monoxide gas on metals, in particular nickel [16]. This gas is traded as an initial material or intermediate in the production laboratories of hydrogen, ammonia, aldehydes, and dyes. This gas interacts with the minerals to form volatile carbonyl compounds [13]. It also affects the nickel, one of the components of the stainless steel alloy, which turns into a volatile material and liberates it from the surface of the metal, leaving gaps in the areas it deposits in the alloy and becomes irresistible to pressure, which leads to the swelling of these tubes Weak mechanical resistance.

1.3.2 Electrochemical corrosion

This type of corrosion occurs due to a real chemical reaction between corrosive medium and the metal or the composition accompanied by transferring electrons between two sites on the surface, one with high electron density and another site with little electron density or between two points, one with a high voltage and another with a low voltage with excises of the electrolyte solution is the corrosive environment or transport medium the medium that transmits electrons [17]. This type of corrosion converts the metal to its ions and dissolves at the anode.

The resulting electrons are transferred to the cathode area where the reaction reactions or electrons are obtained depending on the type of medium. Corrosion reactions in the conductive solutions of the electrodes are always accompanied by passage. So the corrosion is described as electrochemical [18]. In order for the electric current to flow, an elevator, cathode, and electrolytic solution must exist. The difference in voltage is created between the two points. Corrosion, like other chemical reactions, is subject to the laws of Thermodynamics, as it is possible to predict the possibility of corrosion under certain conditions of the amount and type of change in the value of free energy to interact with corrosion and change in free energy (ΔG) linked to the electrochemical cell voltage (E) [18].

$\Delta G = -nfE$; where n is the electron numbers involved in the reaction, "f "is Faraday's constant C.mol-1 (96500), E is Cell potential (Volt). The reaction is likely to occur if the

(ΔE) sign is negative and E positive, and the range of reaction depends on the value of the voltage [19].

The cell voltage for any electrochemical reaction is usually calculated from the Nernest equation. The polar voltage can be measured using a voltmeter voltage meter after assigning it to a reference inert polarity that is placed in the same solution (a pole voltage and not a corrosion voltage) [20]. This voltage is usually referred to as a half-cell voltage because it is a single-pole voltage that can be eroded and hemmed Likewise, there is a voltage of the half-cell pole if the electrode behaves like a cathode and a reduction process occurs [20].

1.3.3 Homogeneous (general) corrosion

This type of corrosion occurs with the same intensity on all parts of the metal surface and usually includes wide areas of the surface [21]. This type of corrosion is the most common form of corrosion and can occur in humid or dry climates, as well as by chemical or electrochemical mechanics [21]. It can be controlled through a good selection of materials and design and by using methods of covering or painting to protect against corrosion, and homogeneous corrosion is the simplest form of corrosion in terms of the ability to determine the amount and speed of its occurrence and can be separated from the file [22].

1.3.4 Local corrosion

This type of corrosion occurs on a certain location of the metal surface more strongly, and its continuation leads to a hole in the surface, and it may lead to holes in the metal if its thickness allows this. This type of corrosion is more effective than smooth and corrosive. Aggravation of the matter [23]. Local erosion can be divided into two types according to the dimensions of the effects that it leaves on the metal surface. First: Local erosion, second: microscopic local erosion and excitability [24].

Small local corrosion is visible only to the microscope. Microscopic local erosion

Intergranular corrosion:

This type of corrosion occurs in metals or polycrystalline alloys. When cooling a molten metal, it begins to freeze at many crystallization nuclei distributed randomly inside the molten metal, and it is formed from those nuclei in different directions by the uniformity of the metal atoms in a specific engineering pattern consistent with the type [25]. The crystallization with which the metal atoms undergo to form what is called grains. The uniformity of the atoms in each bead or crystal and the distances between them are the same in all grains, but as a result of the random distribution of the nuclei of the crystallization, the levels of atoms that fall in the vicinity (between the grains) will not be subject to the system of any of the grains [26]. The atoms are located in the border locations between the granules will differ in their organization and layout arranged and it would take coordinated compromise, these sites are called grain boundaries area, or between granules [26]. It is known that these granular boundaries are more active in interacting with the corrosive medium than the granules themselves [27]. This is because the metal atoms in the region, which are taking an intermediate position between two or more grains, have not taken the position of equilibrium like the rest of the grained atoms, so they are at a higher energy level, which makes them more active in interacting with the eating environment. In addition to this, impurities may collect at borders Granules create an opportunity for the presence of atoms of various other metals in this site, which provides the opportunity for galvanic corrosion to occur at this site. The exacerbation of intergranular corrosion results in the grains dislodging themselves, leaving a rough surface with perceived effectiveness of corrosion [28].

1.3.5 Stress – corrosion cracking

This type of corrosion occurs as a sequence of conjugated action of both mechanical tension and the corrosive medium, and when certain conditions are present the metal cracks, most of the metal alloys are exposed to this type of corrosion [29]. Each alloy has a specific corrosive medium that causes cracks to appear when mechanical stress is placed on it. The stresses that cause such cracks or cracks may be stored in the metal during cold formation and during welding or thermal transactions or when external stress is shed while using the metal or alloy [30]. Cracks can take paths between the grains or across the grains, and cracks may branch and branch through these paths. Among the methods used to reduce this type of wear are [31]:

1. Release or rid the metal structure from the stored stresses with the appropriate thermal treatment.

2. Removing the corrosive medium that causes corrosion from the operating medium.

3. Replacing the alloy with one more resistant to this type of corrosion.

1.3.6 Galvanic corrosion

This type of corrosion occurs when two different metals in their electrochemical activity come into contact with each other or when they are connected through an electrical conductor [32]. As a result of contact with or delivery of the two metals together, each metal acquires a certain electrical voltage, the value of which depends on the type of metal and its location in the galvanic chain and on the concentration of the electrolyte solution in addition to the temperature [33]. A potential difference arises between the two metals, which act as an electric driving force and the passage of an electric current through the electrolytic medium or the corrosive medium [34]. The passage of this electric current results in the erosion of the active and lowest metal in the galvanic chain, as the anode acts for the cell and the oxidation process and the release of electrons, occur while the other metal acts as a cathode. The higher the difference in voltage between the two metals, the greater the chance that galvanic corrosion will occur more quickly [32]. The relative area of contact with the two metals together is of great importance in increasing or decreasing the corrosion velocity. When the area of the less active metal is less than the area of the most active metal, this accelerates the process of corrosion [33].

The occurrence of this type of corrosion can be avoided by placing a buffer between the metal when it calls for the need to use a metal double in a metal stomach. It is preferable to use two closely related metals in their standard voltage so that the difference in voltage created between them is small [35]. The intensity of this corrosion can also be reduced by reducing the area of the most active metal and increasing the area of the other metal [36].

1.3.7 Erosion corrosion (EC)

This type of corrosion occurs when a metal is destroyed by two factors, one chemical and one mechanical, as is the case when a metal surface moves in the corrosive medium [37]. This is since the products of the corrosion process, which can act as a protective layer, if deposited porously, will disappear because of the movement of the metal surface or the corrosive medium [37]. This type of corrosion is common in pump and tube propellers, and this type of wear can be avoided by changing the design or good choice of materials with high resistance to Erosion corrosion. There are two types of EC, which is the corrosive erosion that occurs due to the collision of the bubbles and vaporization of the bubbles at the metal surface, so the high pressure resulting from the bubbles bursting at the surface leads to deformation of the metal surface and the removal of the protective film from the surface [38]. On another metal, this results in the collapse of one or both of the metals together and slipping usually occurs as a consequence of vibrating processes, and the heat produced by friction helps to oxidize the surfaces of the metal, and it soon causes the removal of the formed oxide layer continuously, so the active metal is exposed to friction and heat due to corrosion caused by them [39].

This type of corrosion occurs with the first two mechanics: (Wear-Oxidation theory), which is based on the principle that cold fusion or fusion occurs at the interface between the metal surfaces under pressure and through subsequent relative movement and that these contact points will be directly oxidized [40]. This process is then repeated, producing a loss of the metal and collecting residue Oxides: The assumptions of this theory are based on the principle that frictional wear and tear causes a risk followed by oxidation as a secondary effect [40].

The second theory, known as Wear – Oxidation, depends on the hypothesis that most mineral surfaces are protected from atmospheric oxidation with a thick adherent oxide layer. When these minerals are placed in contact under the influence of load and exposed to frequent relative motion, the oxide layer will crumble at high protrusions and produce oxide debris. In this theory, it is assumed that the exposed metal will be oxidized, and the process repeated [41]. This theory relies mainly on the principle of accelerating oxidative stress attributed to frictional effects. This kind of wear can be reduced by choosing more hard materials or by adjusting the design to ensure that friction operations are avoided. Use of grease may be used to facilitate slipping or to increase surface roughness to the extent that slipping becomes impossible. EC when occurring is characterized by the formation of grooves, ripples, heights or depressions on the surface of the metal. EC was made of tin bronze (Cu-10Sn-Zn) and takes a shape whose direction is the direction of movement of the medium. The following figure shows the undulating nature of the surface of the propeller of an injury pump used for Slurry pumping contains on 5-10% H_2SO_4. The failure occurred after six months of use in this medium [42].

1.3.8 Crevice corrosion

The cracks are usually found in certain locations of the metal structure, as is the case with seals, fittings, nuts, etc., and cracks may form due to the deposition of dirt, corrosion, and scratches in the coating layers [43]. This type of corrosion occurs because the conditions inside the cracks change over time significantly, compared to other areas near the cracks on the metal surface. Erosion may be caused by a change in the pH inside the cracks or a lack of oxygen inside the cracks [44]. Scientists also attribute the cause of this corrosion to the lack of corrosion inhibitors inside the cracks or the accumulation of certain types of ions inside the cracks. It is not necessary for this type of corrosion to occur in all circumstances when a metal meets a corrosive medium, as there are certain metals or alloys that are more susceptible to this type of corrosion than others [45]. It affects metals or alloys that depend on the presence of air to create a shielding oxide layer on its surfaces, such as stainless steel and titanium. The resistance of these materials to corrosion cracks can be improved by adding other casting elements. Also, resistance to this type of corrosion can be improved with good design so that the mechanism for creating these cracks and continuous maintenance is useful in making the surfaces clean continuously and preventing the accumulation of dirt on them [44].

1.3.9 Pitting corrosion (PC)

The Pitting corrosion is the result of a sharp local erosive attack, and the impact of the click may be severe when its depth increases and leads to the formation of holes in the metal or alloy, and then click on the metal surface is formed at different depths and the shape of the PC is a reason for its continuation and growth [45]. To reduce this type of erosion, the surfaces must be clean, homogeneous, and polished. Usually, the process of PC growth is slow, which may take months or years until the PC becomes visible with the eye, but it always causes metal collapses without warning, and the non-click size and the small amount of metal that is needed dissolving it for a click makes it difficult to detect this type of wear in its first boiler [46]. A good selection of materials, design, and manufacture is the best and safest method to avoid this type of corrosion. Halogens ions, in particular chloride ions, have the ability to remove the negative layer from the metal or the alloy by breaking down the light protective layer of oxide that has formed on the surface, and the chloride ions remove the negativity by penetrating the protective layer in certain areas causing PC [47]. Chloride ions may do their work by competing with oxygen and a hydroxide ion to bond to the surface of the metal and break down the oxide layer [48]. There is a general agreement that the spread of clicking in a metallic substance includes the melting of the metal and high acid events at the bottom of the tap due to the hydrolysis of the dissolved ions.

The process of removing the negativity, the oxidation layer breakdown, and the occurrence of click erosion on the surface of a metal or alloy occurs when an electrical voltage is reached in the negative region. This voltage is called the critical voltage of the PC [49]. The value of this voltage can be obtained from the polarization curves obtained using the static voltage meter. The critical voltage value decreases by increasing the chloride ion concentration, by reducing the pH and increasing the temperature. The critical voltage value can be raised to PC by adding some salts, such as sodium nitrate, above sodium chlorate, and salts containing sulfate ions, hydroxide, chlorates, carbonates, or nitrates [47].

1.3.10 Exfoliation corrosion

It means what happens without a metal surface (or below the surface of an alloy), as it starts at the clean surface but spreads below it and differs from fractional corrosion in that attacking the metal's corrosion is taking a soluble metal image, and it can be reduced by conducting the necessary thermal treatments or Casting with other metal [50].

1.3.11 Selective leaching

It is intended to melt one component of an alloy while keeping the other components as is the case when extracting silver from

Some silver ingots with gold, and thus the alloy, after removing one of its components, will have a porous nature and weak mechanical properties. Examples of this type of corrosion include Dezincification, Graphitization, Dealuminumification, and Decobaltification [48,51].

1.3.12 Nonmetallic corrosion

Ceramics or ceramics consist of high levels of silica and is used in lining the walls of laboratories, laboratories and factory floors in addition to lining the equipment from the inside to resist the impact of cold and hot acids, it has a very high resistance to various types of organic and inorganic acids as it perfectly resists erosion but is weak for fluctuations in temperature and shocks and damages quickly in the presence of fluoride ion and hydrofluoric acid which dissolves silicates according to the following equation, and is also affected by the rules with high concentrations and at high temperatures [51].

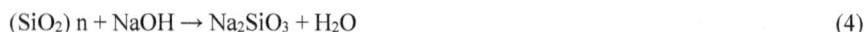

$$(SiO_2) n + HF \rightarrow SiF_4 + H_2O \tag{3}$$

$$(SiO_2) n + NaOH \rightarrow Na_2SiO_3 + H_2O \tag{4}$$

1.3   Corrosion of cement

Cement is a major building material in the industry and the construction of various industrial buildings [53]. There are many types. It has varying resistance to biting climates. The use of the right type of cement for the appropriate medium gives excellent resistance and uses the ordinary type of cement in abundance for its cheapness and ease of use. It may be coated with a layer of porcelain, rubber or plastic to increase its resistance to rodent climates [54]. The water effects are the effect of the soft water: the soft water affects the cement, forming calcium hydroxide and salt crystals containing several water molecules as crystalline water according to the formula:

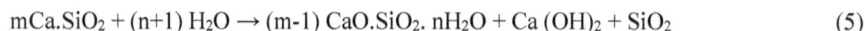

$$mCa.SiO_2 + (n+1) H_2O \rightarrow (m-1) CaO.SiO_2. nH_2O + Ca (OH)_2 + SiO_2 \tag{5}$$

1.5 Corrosion of organic materials

These materials include polymers, dyes, wood and textiles.

The long molecules of the polymers and convert them into smaller molecules with different physical specifications than the long molecules, as the polymers are distinguished from each other by the degree of polymerization and it is known that the specifications of the high density polymer are better than the low density and that the breakdown of the long polymers particles converts it from.

The most important distortions to which the polymers are exposed are also deformation due to electrical currents, continuous mechanical movements and even high sound waves [55]. As for corrosion due to chemicals, it is the oxidation of the polymers and breaks them with a strong oxidizing medium or by the influence of oxygen, as the polymers are decomposed by the influence of acidic and basic waters. Pure or electrochemical, but they are caused by vital activity, it is known that many microorganisms, including algae, and bacteria, are present in the air, water and inside the soil, and that these microorganisms are vital activities that feed on organic matter and for inorganic and acid production or bases that create harmful atmospheres (corrosion surroundings) for the metal or the structure on which it grows. Wood, for example, is made of cellulosic fibres that are linked with each other by bonding materials that are the lignan [56]. Large cooling towers used in oil refining cooling systems and production plants. The growth of algae on the Lignin material loses the strong wood and materials that bond to cellulosic fibres and thus lead to its breakdown [57].

The breakdown of wood, the breakdown of the structure, or its weak mechanical resistance, resulting in damage to the cooling towers. To protect from this, the wood used in building the cooling towers is covered in or coated with materials that prevent algae from growing (dead materials) or using a biocide within the water cycle [52]. There are other microorganisms that cause corrosion of iron and other minerals we see inside oil or water pipelines, which are the anaerobic bacteria that feed on sulphates and turn them into sulfide. It is known that the presence of deposits or dirt on the surface of the metal prevents air, oxygen or the external solution from reaching the surface of the metal, which provides an anaerobic atmosphere very suitable for the work of this bacterium as it produces sulphur dioxide [46].

The solubility of sulphur dioxide in the water produces sulfuric acid and converts the sulfide into hydrogen sulfide, which leads to a decrease in the pH and acid corrosion on the surface of the metal. The product of the reactions is the oxygen gas that causes increased cathode polarization, that is, the withdrawal of electrons from the cathode, which stimulates anode reactions and increases the speed of corrosion [57]. To get rid of anaerobic bacteria inside the tubes, we use appropriate methods to prevent the accumulation of sediments or dirt and the use of microorganisms' killers in the restrooms of closed systems. As for aerobic bacteria, they work in a normal medium (not under deposits or corrosion products), such as thiobacteria thioxides, which have the ability to oxidize sulphur or sulfide to sulfuric acid. These bacteria, which work in an acid medium, can produce sulfuric acid with a concentration of up to 5%. These bacteria are found in many sulphur productions fields, oil fields, and water drainage pipes [58].

The factors affecting the rate of corrosion rate are the most common process in the industry, because other types of corrosion can be expected and avoided more easily, and electrochemical corrosion processes require the availability of three conditions for their occurrence: the metal obtained by corrosion, the medium that causes corrosion, and a vector of electrification with two types of electrons [52]. And ions. There are two types of factors that affect the speed of corrosion: crystalline forms and the uniformity and uniformity of crystals affect the resistance of the ocean. During the manufacture of alloys by mixing specific proportions of casting elements, they undergo various processes, including smelting, to convert to a liquid state, and cooling to shift to different crystalline forms [46]. The cooling speed affects the crystalline shape, its concentration, and the amount of amorphous materials that solidify with it, and thus the quality and properties of the alloy formed. The occurrence of a defect in one atom of 10^{10} generates a significant and visible effect on the properties of the metal, and a large number of types of defects of alloys have been identified. These defects like local defect. This defect results from the loss of an atom within the cell unit, resulting in a vacuum or penetration of a large atom or a small ion compared to the size of the original atoms forming the crystal leading to the presence of a broken atom. Linear imbalance, which is caused by the presence of an extra half-plane of atoms within the crystalline structure. Surface imbalance, generated by the agglutination of crystals during their conversion from a liquid to a solid state in the opposite direction 180 from what is assumed, causing a crystal shift or a mirror image of the other crystal. Dysregulation, caused by an increase or decrease in an entire crystalline level within the crystalline structure [58]. As for the defect of solid solutions, it occurs due to the deposition of impurities within the crystal structure of the same size as atoms, but of minerals. Different, so they are solid prosthetic solutions or broken solid solutions. Amorphous in solid, it occurs due to the freezing of amorphous materials at the crystalline boundaries of the same elements that make up the crystal, but in different proportions and amorphous. During the formation and installation process, the metal is also exposed to various forms of stress that cause a change in the properties of the metal in general, including the engineering shape and the method of fluid or fluid flow: the speed of corrosion is faster in the reverse areas due to mechanical stresses in that region and finally the metal is exposed to air corrosion during Long storage stage or during transportation from the place of origin. To the place of the monument due to damp weather conditions or scratched abrasion conditions, therefore, care should be taken to use air corrosion inhibitors.

1.6   Environment factors

1.6.1   Effect of oxygen and oxidants

There are several factors related to the corrosion medium affecting the rate of corrosion rate, including: the effect of oxygen on increasing the speed of corrosion to the state in which the surface of the metal has completely covered the buffer oxide layer from the surrounding where a severe decrease in the rate of corrosion occurs with an increase in the oxidizer concentration and the condition remains so throughout the period In which the metal is in a negative state and then the speed of corrosion increases again in the post-negative phase (due to the formation of several layers of oxide which will be fragile or perhaps because of changes in the electrode voltage values of the metal and oxidation reactions of another type occur. In all types of erosions Admonishes O oxygen is the main element in the occurrence of the process of electrochemical corrosion except for the corrosion caused by hydrogen. The movement of the ocean affects negatively or positively on the speed of corrosion. The erosion caused by the movement of the medium, as the decrease in the velocity of the medium reduces the speed of corrosion [51].

1.6.2   Effect of pH

Minerals differ in their acid and basic properties in the periodic table, as are the elements in the group.

The first has basal properties, and the more we turn to the right of the periodic table, the more acidic properties are obtained, and we get acid oxides at the seventh group (i.e. acid resistance). Most of the minerals we deal with have amphitheatre properties and fall within the transitional elements.

1.6.2   Effect of anions and cations

We have previously discussed in different places the effect of the presence of negative chloride ions in the solution that causes. In the events of focal erosion and cracks corrosion, since chloride salts are water-soluble salts and as chlorides, therefore chloride ions increase corrosion. As for nitrate ions (NO_3^-), their presence is positive, even though they are dissolved salts, because nitrate ions are oxidizing agents that adsorb to the surface, forming an insulating layer that prevents corrosion in static media. In the moving media, this layer is poorly adsorbed and does not stand up, but it causes and helps to create a protective oxide layer that reduces the rate of corrosion (58). The presence of positive chromium ions in the vicinity of the corrosion of iron is positive. If this ion adsorbed on the surface and is more effective, there will not be corrosion on the iron site near this region in which the adsorption of chromium atoms occurred. As for the presence of positive copper ions with iron, it is negative because copper is less active than iron. If the adsorbed copper, it will behave like a cathode with respect to iron, so the area near the adsorption zone will be eroded. Dissolved gases also have effects on corrosion velocity, and the following figure shows the corrosion velocity relationship in percentage of sulfuric acid for corrosion of iron. It is observed that the lowest speed we obtain at concentrations is 80-100% and the highest corrosion is at the 40% boundary, which is the same areas where the electrical conductivity of the sulfuric acid solution It has the greatest value due to the occurrence of various forms of bonding between water, acid and SO_3 [53].

1.7 Anti-corrosion methods

Corrosion cannot be stopped or prevented from happening altogether, but there are several ways to prevent it or reduce the speed of its occurrence to the lowest speed, which prolongs the life of the stomach or machine as long as possible and work with efficiency similar to design efficiency. The process of prevention or protection from corrosion is done in several ways: Metal and surrounding variants, good stomach design, use of corrosion inhibitors (inhibitors), and electrical protection (cathodic and anodic) and coating or coverage. The corrosion inhibition mechanism is the absorption of the barrier on the metal surface [49].

The Various determinants influence the adsorption method, such as the nature of the mineral, the chemical composition of the organic inhibitor, and the aggressive electrolyte. Some of the critical elements that make a contribution to the efficiency of corrosion inhibition are: a) the size of the organic molecule, b) the aromatic odors and/or bonding, c) the length of the carbon chain, d) the bonding power of the mineral layer, e) the kind of the connected atoms inside the molecular system, f) the potential of the layer to end up compressed or bound together, g) the ability to shape a complex atom as a solid within the metal mesh, h) and adequate melting within the environment. Most of the efficient inhibitors are organic compounds that contain atoms that have N, S, or O atoms in their structures, practical electrical groups, and electrons in double bonded. The inhibitory action of these natural compounds may be attributed to their interactions with the surface of the metal via the adsorption process. Adsorption of organic compounds on the surface of the metal is often caused by: metallic electrophoresis on the charged particles; the transaction of the single pair electrons of heterogeneous atoms with the unoccupied orbital of the metals; or a mixture of some previous mechanisms. Thus, the absorption of inhibitors depends first at the steel and its surface charge. For all the factors stated above, it is pretty clear that the damping mechanism must be carefully studied because it affords a mandatory amount of data that ought to be taken into account when running in specific corrosion inhibition in specific environments [59].

The purpose of inhibitors is to form a single bulkhead or several molecular layers resistant to acids intervention. This preventive measure is frequently chaperoned with analytic and/or physical adsorption that includes contrast in the charge of the absorbent material and the convey of the charge from one stage to another. Heterogeneous compounds that do not contain sulphur with or without nitrogen in the presence of various alternatives are highly efficient corrosion inhibitors. The thiophene and hydrazine derivatives are also special compounds to prevent erosion of metals in acidic solutions. Inorganic substances such as phosphates, chromates, silicate dichromate, borates, tungstates, molybdates, and arsenic are effective as inhibitors of mineral corrosion.

1.7.1 The green impediments for corrosion

The corrosion curbs which extracted from the green plants are biodegradable and do not contain heavy metals or other toxic compounds. Some research groups have reported the successful use of naturally occurring materials to prevent mineral corrosion in both the acidic and alkaline environment.

The structure of the active ingredient is taken into account when green inhibitors are working. Many researchers have hypothesized several theories to explain the mechanism of their effect. It should be noted that the active ingredient derived from natural inhibitors varies from plant type to another, but their compositions are closely related to organic coordination [58].

Phenols and polyphenols are other important classes of natural compounds found in plants. The corrosion inhibiting action of these compounds is mainly due to the presence of aromatic hydroxyl groups. This was explained again by the ability of heterogeneous atoms to coordinate bonds with vacant orbits of the metal by donating electrons. Also, interactions with loops that have bonding bonds, π electrons, enable the molecules to be absorbed onto the surface of the metal. A study established the inhibition of carbon steel corrosion using olive leaf extract. In this investigation, dehydrated olive leaves were boiled in water for three hours, then after evaporation and filtration, the solid residues were used to develop the required solutions for the inhibitors. The inhibition phenomenon towards acid corrosion of steel is attributed to absorption by individual pairs of electronic oxygen atoms that form a layer and act as a barrier between the surface of the steel and the attacking solution. The aqueous extract of leaves (*Olea europaea L.*) contains polyphenols with antioxidant capacity. On the same approach, components in oleuropein and hydroxytyrosol other existing polyphenols were identified that include tyrosol, olerobene glycone, and gallic acid. Hydroxytyrosol may be the main responsible for the inhibition process however, there was no empirical evidence to confirm this hypothesis about the mechanism of inhibition action and the activity of individual compounds at work has not been verified.

1.7.2 Determination of green corrosion inhibitors based on ionic fluids

Currently, increasing global environmental precaution and stringent environmental protocols restrict the synthesis and use of hazardous conventional volatile corrosion inhibitors. Therefore, there is a vital need for improvement in synthetic and engineering chemistry with environmentally friendly materials. The use of alternative synthetic strategies toward green synthesis has become necessary as multi-component reactions with ultrasound (sonochemical) synthesis and microwave irradiation. Therefore, there's an urgent want to enhance structural and engineering chemistry with environmentally friendly substances. The use of alternative synthetic strategies in the direction of non-toxic synthesis has grown to be vital as reactions associated with ultrasound (sonochemical) synthesis and microwave radiation. Therefore, scientists had been directed to develop inhibitors from plant extracts which might be characterized by using the non-toxic nature to act as green corrosion inhibitors [57]. However, extracting plant extracts is cumbersome, costly, time-ingesting and entails relatively big amounts of natural solvents that may negatively affect the surroundings and dwelling organisms. High temperatures can also purpose the decay of the lively ingredients and reduce the performance of the relative inhibition. For these considerations, the use of extracts as corrosion inhibitors for ferrous and non-ferrous metals is also limited. Therefore, there is a need to develop green inhibitors through proper synthesis design, which can be achieved using cheap and environmentally friendly products. This is the beginning of green chemistry such as ionic fluids that are environmentally friendly and sustainable solvents consisting of ions that can replace a wide spectrum of inorganic and organic compounds. Ionic fluids follow the principles of green chemistry recommended by Anastas and Warner [59].

Lately, many studies focus on raw extracts, and the development of these inhibitors requires the binding of the inhibitor's efficiency to its molecular properties through the interaction between inhibitors and the mineral surface and describing the adsorption behavior of ionic fluids on the surface of the metal. This is achieved through the theory of functional efficiency and is one of the most important methods used in theoretical chemistry to inhibit corrosion and to accurately predict the efficiency of inhibition based on the calculations provided by mechanical information about the reactions of mineral inhibitors and can be made to inhibitors before synthesis [60].

Inula plant extracts as an anticorrosive agent to cover iron samples and protect them, as this plant is present in the coastal environment and is inexpensive and can be obtained easily and easily [61,62]. It is a perennial tree plant, its base is woody, its strong camphor aroma is strong and jet, its flowers are yellow, its leaves are spear-shaped and hairy with a sticky feel. Moderate is most suitable for the growth of herbaceous herb, so we see it extensively on the outskirts of the Mediterranean basin and regions of North Africa. In addition to this plant has good antioxidant properties that define the best conditions for extract (Inula, seawater) as it was (PH = 7.1, time = 24 hours, concentration = 10%) The studied iron samples were treated with extract (mud, seawater) under the same specific conditions previously and studied the change in the rate of corrosion speed of these samples and the effectiveness of the inhibitor effect of these samples [62,63]. With untreated samples, the values of the wear rate were pre-and after treatment as follows: the speed of the rate of erosion of iron samples (before the treatment with the extract) within 120 days.

The effect of adding saffron extract on the corrosion of α-brass mineral in nitric acid by chemical method, electrochemical method, electrochemical impedance spectroscopy, dynamic voltage polarization, and electrochemical frequency modulation technique [64]. As the results proved that the inhibitor efficiency increased with increasing concentration inhibitor, but decreased with increasing temperature, the extract works by forming an adsorbent layer on the surface of the metal that protects it from corrosion of the oxidizing environment, the adsorption of the extract on the surface of the alloy α- Brass naturally and follows the theory that an adsorption isotherm. The effect of temperature on corrosion of the metal has been studied in nitric acid at a concentration of 1 Mole. Dynamic voltage depolarization studies have shown that safflower extract is an excellent inhibitor in keeping metallic ingots understudy from corrosion [64].

Organic compounds that contain negatively charged functional groups and electrons are in bilateral bonds [65]. The triple is usually considered good inhibitors of acid wear. In recent years, many phosphine salts and oxides have been prepared, but most of them have been directed to the biological and medical side, especially as they can form stable complexes. The surface of the minerals can be considered as inhibitors as possible [66]. Note that phosphorous has better effectiveness than oxygen in terms of inhibition. These compounds have good solubility in distilled water, as it was studied after dissolving it in hot distilled water. Some studies have recorded the efficiency of green tea on the performance of the 316-type steel used in the compensatory process for affected human body parts. These studies have shown positive results indicating the improvement of the performance of the alloy inside the body [67].

1.7.3 Corrosion suppressions from the biological waste

The significance of achieving biological waste in the context of the circular economy is crucial. There are scant scientific papers in dealing with natural waste and bio-waste, which is usually disposed of in the environment. Some investigations have focused on pectin arisen from citrus peel; others on chitosan extracted from seafood waste [68]. Fresh leaves are used from plants grown for fruits (bananas, cane, and watermelon). It started with peel, seeds and melon peel, as well as using tomatoes derived from by-products as pectin sources [69].

Water is widely used in heat exchange to remove unwanted heat in some industrial operating systems such as petrochemical industries and others. As a result of the growth and growth of the population, there is a trend to reduce water use through reuse [70]. But the warm water out of the cooling process contains dissolved salts, suspended matter, dissolved and suspended organic matter, gases, and microorganisms. This may lead to corrosion of the minerals and the formation of limestone crusts deposited inside the pipes of the cooling systems, which results in partial or total blockage of these pipes, thus causing severe economic loss. Some research use the extracts of several natural herbs like *Solenostemma argel* (*S. argel*) and oregano, and seaweed (green and red) to reduce the formation of crusts and corrosion of steel [71]. Previous results have confirmed the efficiency of herbal extracts (*S. argel* and oregano) and algae extract (green and red) in inhibiting the formation of crusts and inhibiting corrosion of steel. The results also showed an increase in the efficiency of herbal extracts (barn and oregano) over algae extract (green and red) in inhibiting the formation of crusts and inhibiting the corrosion of steel. The results also showed an increase in the efficiency of herbal extract from the other previous inhibitors. A comparison of the efficiency of using the optimum concentration of *S. argel* extract and 10% of a chemical mixture consisting of (11.1 ml sodium triphosphate + 11.1 ml carb hydrazide) in inhibiting crust formation and inhibiting the corrosion process of steel used in petrochemicals [72]. The results indicated an increase in the efficiency of the optimum concentration of kelp extract from the chemical mixture in inhibiting the formation of scales and inhibiting the erosion of steel. This method is inexpensive and safe for the environment, and it consists of available natural materials and is not complicated to manufacture and plays the same role as chemicals imported from abroad, which will lead to saving money and establishing factories. Testing of red apple scales and pomegranate as a natural environmentally friendly inhibitor for copper-zinc alloy in acidic media (sulfuric acid and hydrochloric acid) and at different temperatures. Various spectral techniques (infrared spectroscopy and gas-mass spectrometry chromatography) have been applied to know and detect the natural organic compounds present in the crusts that are responsible for inhibiting the alloy erosion because of the various atoms that have a high electron density necessary for the inhibition process [72]. Also, the alloy's surface was studied before and after adding scaling extracts using the scanning electron microscope and X-ray diffraction, as these techniques demonstrated the effectiveness of these inhibitors. Through the adsorption calculations and the corrosion values obtained from the polarization study, the inhibition efficiencies were calculated as the highest inhibition efficacy of red apples was 93% and Pomegranate 91% [73]. These values proved that the natural compounds present in the crusts are highly effective compared to the prepared organic inhibitors that are of high and costly toxic and difficult to prepare [73].

Conclusion

The organic corrosion inhibitors can be also immobilized and used as biocorrosion inhibitors. Besides, natural corrosion inhibitors can be activated and used as biological corrosion inhibitors. In general, the corrosion inhibition mechanism is adsorption on the metal surface. The adsorption process is affected by various factors such as the nature and charge of the metal, the chemical composition of the organic inhibitor (functional groups, aromatic, potential aseptic effects) and the type of electrolyte that causes corrosion. Some of the critical elements that contribute to the effectiveness of corrosion inhibition are the dimensions of the organic molecule, the length of the carbon chain, the binding energy of the mineral substrate, the type and amount of atoms bonded in the molecule, the ability to become a compact or bonded layer, and the ability to form a complex with the atom as a solid in the network Mineral, good solubility in the environment. Natural products contain various organic compounds such as aldehyde, ketogenic, amides, amino and organic acids, and they are often known for their inhibitory effect.

The future research in the field might be addressed in the directions of a broader circular economy background with a careful evaluation of the starting source, coming to terms with the value of the source itself.

References

[1] C.D. Taylor, Corrosion informatics: an integrated approach to modelling corrosion. Corros. Eng. Sci. Technol. 50(7) (2015) 490–508. https://doi.org/10.1179/1743278215Y.0000000012

[2] S. Syed, Atmospheric corrosion of materials, Emirates J. Eng. Res. 11(1) (2006) 1–24.

[3] M.G. Fontana, Corrosion Engineering. Tata McGraw-Hill Education, 2005.

[4] B. Valdez, M. Schorr, R. Zlatev, M. Carrillo, M. Stoytcheva, L. Alvarez, Corrosion control in industry, Environ. Ind. Corros. Pract. Theor. Asp. 2012. https://doi.org/10.5772/51987

[5] P.A. Schweitzer, Fundamentals of Corrosion: Mechanisms, Causes, and Preventative Methods. CRC press, 2009. https://doi.org/10.1201/9781420067712

[6] G.J. Jorgensen, K.M. Terwilliger, J.A. DelCueto, S.H. Glick, M.D. Kempe, J.W. Pankow, Moisture transport, adhesion, and corrosion protection of PV module packaging materials. Sol. Ener. Mater. Sol. Cells. 90(16) (2006) 2739–75. https://doi.org/10.1016/j.solmat.2006.04.003

[7] D. Landolt, Corrosion and Surface Chemistry of Metals. EPFL press, 2007.

[8] L. Tushinsky, I. Kovensky, A. Plokhov, V. Sindeyev, P. Reshedko, Coated Metal: Structure and Properties of Metal-Coating Compositions. Springer Science & Business Media, 2013.

[9] P.A. Schweitzer, Fundamentals of Metallic Corrosion: Atmospheric and Media Corrosion of Metals. CRC press, 2006. https://doi.org/10.1201/9780849382444

[10] H.A.A. Al-Mazeedi, R.A. Cottis, A practical evaluation of electrochemical noise parameters as indicators of corrosion type. Electrochim. Acta. 49(17–18) (2004) 2787–93. https://doi.org/10.1016/j.electacta.2004.01.040

[11] E. McCafferty, Introduction to Corrosion Science. Springer Science & Business Media, 2010. https://doi.org/10.1007/978-1-4419-0455-3

[12] G.B. Darband, M. Aliofkhazraei, S. Khorsand, S. Sokhanvar, A. Kaboli, Science and engineering of superhydrophobic surfaces: Review of corrosion resistance, chemical and mechanical stability, Arab J. Chem. 13(1) (2020) 1763–802. https://doi.org/10.1016/j.arabjc.2018.01.013

[13] M.A. Malik, M.A. Hashim, F. Nabi, S.A. Al-Thabaiti, Z. Khan. Anti-corrosion ability of surfactants: A review. Int. J. Electrochem. Sci.6(6) (2011) 1927–48.

[14] C.A. Loto, Microbiological corrosion: mechanism, control and impact—A review. Int. J. Adv. Manuf. Technol. 92(9–12) (2017) 4241–52.

https://doi.org/10.1007/s00170-017-0494-8

[15] R. Sydney, E. Esfandi, S. Surapaneni, Control concrete sewer corrosion via the crown spray process, Water Environ. Res. 68(3) (1996) 338–47. https://doi.org/10.2175/106143096X127785

[16] D.R. McIntyre, G.T. Burstein, A. Vossen, Effect of carbon monoxide on the electrooxidation of hydrogen by tungsten carbide, J. Power Sources. 107(1) (2002) 67–73. https://doi.org/10.1016/S0378-7753(01)00987-9

[17] N. Perez, Electrochemistry and Corrosion Science. Vol. 412. Springer; 2004. https://doi.org/10.1007/b118420

[18] B.E. Rani, B.B.J. Basu. Green inhibitors for corrosion protection of metals and alloys: An overview. Int. J. Corros. Article ID 380217 (2012). https://doi.org/10.1155/2012/380217

[19] M. Pourbaix, Lectures on Electrochemical Corrosion. Springer Science & Business Media, 2012.

[20] M.C. Kennedy, Electrochemical study of the metal-chalcogenide glasses: fundamentals and applications in ion selective electrodes. 1985.

[21] M. Malinina, T. Sammi, M.M. Gasik, Corrosion resistance of homogeneous and FGM coatings, Mater. Sci. Forum, 492-493 (2005) 305–310. https://doi.org/10.4028/www.scientific.net/MSF.492-493.305

[22] J. Rodriguez, L.M. Ortega, J. Casal, J.M. Diez, Corrosion of reinforcement and service life of concrete structures. Durab. Build. Mater. Components 7 (2018) 1-117.

[23] L.P. Huang, K.H. Chen, S. Li, M. Song, Influence of high-temperature pre-precipitation on local corrosion behaviors of Al–Zn–Mg alloy, Scr. Mater. 6(4) (2007) 305–8. https://doi.org/10.1016/j.scriptamat.2006.09.028

[24] G. Yang, L. Ying, L. Haichao, Experimental studies on the local corrosion of low alloy steels in 3.5% NaCl, Corros. Sci. 43(3) (2001) 397–411. https://doi.org/10.1016/S0010-938X(00)00090-1

[25] G. Palumbo, K.T. Aust, Structure-dependence of intergranular corrosion in high purity nickel, Acta Metall. Mater. 38(11) (1990) 2343–52. https://doi.org/10.1016/0956-7151(90)90101-L

[26] K. Holmberg, A. Matthews, Coatings Tribology: Properties, Mechanisms,

Techniques and Applications in Surface Engineering, Elsevier, 2009.

[27] M. De Keersmaecker, M. Dowsett, M. Adriaens, How to preserve lead artifacts for future generations, in: Chemical Interactions Between Cultural Artefacts and Indoor Environment. ACCO, 2018, pp. 215–44.

[28] M. Matula, L. Hyspecka, M. Svoboda, V. Vodarek, C. Dagbert, J. Galland, Intergranular corrosion of AISI 316L steel, Mater. Charact. 46(2–3) (2001) 203–10. https://doi.org/10.1016/S1044-5803(01)00125-5

[29] R.H. Jones, Stress-Corrosion Cracking, Materials Performance and Evaluation. ASM international, 2017. https://doi.org/10.31399/asm.tb.sccmpe2.9781627082662

[30] B. Davo, A. Conde, J.J. De Damborenea, Inhibition of stress corrosion cracking of alloy AA8090 T-8171 by addition of rare earth salts, Corros. Sci. 47(5) (2005) 1227–37. https://doi.org/10.1016/j.corsci.2004.07.028

[31] W. Rostoker, J.R. Dvorak, Interpretation of metallographic structures. Elsevier 2012.

[32] L.H. Hihara, R.M. Latanision, Galvanic corrosion of aluminum-matrix composites, Corros. 48(7) (1992) 546–52. https://doi.org/10.5006/1.3315972

[33] X.G. Zhang, Galvanic corrosion, Uhlig's Corrosion Handbook 51 (2011) 123. https://doi.org/10.1002/9780470872864.ch10

[34] N.M. Taher, A.S. Al Jabab, Galvanic corrosion behavior of implant suprastructure dental alloys. Dent. Mater. 19(1) (2003) 54–9. https://doi.org/10.1016/S0109-5641(02)00008-8

[35] W.-T. Tsai, J.-R. Chen, Galvanic corrosion between the constituent phases in duplex stainless steel, Corros. Sci. 49(9) (2007) 3659–68. https://doi.org/10.1016/j.corsci.2007.03.035

[36] M. Fripp, Z. Walton, Degradable metal for use in a fully dissolvable frac plug, Offshore Technology Conference, 2016. https://doi.org/10.4043/27187-MS

[37] R.C. Barik, J.A. Wharton, R.J.K. Wood, K.R. Stokes, R.L. Jones, Corrosion, erosion and erosion–corrosion performance of plasma electrolytic oxidation (PEO) deposited Al_2O_3 coatings, Surf. Coatings Technol. 99(2–3) (2005) 158–67. https://doi.org/10.1016/j.surfcoat.2004.09.038

[38] R.J.K. Wood, Erosion–corrosion interactions and their effect on marine and

offshore materials, Wear 261(9) (2006) 1012–23.
https://doi.org/10.1016/j.wear.2006.03.033

[39] R. Malka, S. Nešić, D.A. Gulino, Erosion–corrosion and synergistic effects in
disturbed liquid-particle flow. Wear 262(7–8) (2007) 791–9.
https://doi.org/10.1016/j.wear.2006.08.029

[40] R.W. Bruce, Handbook of Lubrication and Tribology, v II: Theory and Design.
CRC press, 2012. https://doi.org/10.1201/b12265

[41] P. Pedeferri, Erosion-Corrosion and Fretting. In: Corrosion Science and
Engineering. Springer, 2018. pp. 313–25. https://doi.org/10.1007/978-3-319-97625-
9_16

[42] E. Huttunen-Saarivirta, F.H. Stott, V. Rohr, M. Schütze, Erosion–oxidation
behaviour of pack-aluminized 9% chromium steel under fluidized-bed conditions at
elevated temperature, Corros. Sci. 49(7) (2007) 2844–65.
https://doi.org/10.1016/j.corsci.2006.12.024

[43] A.J. Sedriks, Corrosion of Stainless Steel, second ed., Wiley InterScience, 1996.
https://doi.org/10.5006/1.3279917

[44] A.K. Mishra, G.S. Frankel, Crevice corrosion repassivation of alloy 22 in
aggressive environments, Corros. 64(11) (2008) 836–44.

[45] Q. Hu, G. Zhang, Y. Qiu, X. Guo, The crevice corrosion behaviour of stainless
steel in sodium chloride solution, Corros. Sci. 53(12) (2011) 4065–72.
https://doi.org/10.1016/j.corsci.2011.08.012

[46] J. Gray, B. Luan, Protective coatings on magnesium and its alloys—a critical
review, J. Alloys Compd. 336(1–2) (2002) 88–113. https://doi.org/10.1016/S0925-
8388(01)01899-0

[47] L.T. Popoola, A.S. Grema, G.K. Latinwo, B. Gutti, A..S Balogun, Corrosion
problems during oil and gas production and its mitigation, Int. J. Ind. Chem. 4(1)
(2013) 35. https://doi.org/10.1186/2228-5547-4-35

[48] A. Jahan, M.Y. Ismail, S.M. Sapuan, F. Mustapha, Material screening and
choosing methods–a review, Mater. Des. 31(2) (2010) 696–705.
https://doi.org/10.1016/j.matdes.2009.08.013

[49] L.Y. Ljungberg, Materials selection and design for development of sustainable

products, Mater. Des. 28(2) (2007) 466–79.
https://doi.org/10.1016/j.matdes.2005.09.006

[50] M. Keddam, C. Kuntz, H. Takenouti, D. Schustert, D. Zuili, Exfoliation corrosion of aluminium alloys examined by electrode impedance, Electrochim. Acta. 42(1) (1997) 87–97. https://doi.org/10.1016/0013-4686(96)00170-3

[51] Z. Yang, M. Rui-Lin, N. Wang-Dong, W. Hui, Selective leaching of base metals from copper smelter slag. Hydrometallurgy 103(1–4) (2010) 25–9.
https://doi.org/10.1016/j.hydromet.2010.02.009

[52] D.L. Galloway, Nonmetallic corrosion-resistant enclosure for electrical apparatus, Google Patents, US3599134A, 1971.

[53] D.-Y. Oh, T. Noguchi, R. Kitagaki, W.-J. Park, CO_2 emission reduction by reuse of building material waste in the Japanese cement industry. Renew. Sustain. Energy Rev. 38 (2014) 796–810. https://doi.org/10.1016/j.rser.2014.07.036

[54] H.R. Rezaie, L. Bakhtiari, A. Öchsner, Biomaterials and their Applications. Springer, 2015.

[55] M. Kliškić, J. Radošević, S. Gudić, V. Katalinić, Aqueous extract of Rosmarinus officinalis L. as inhibitor of Al–Mg alloy corrosion in chloride solution, J. Appl. Electrochem. 30(7) (2000) 823–30. https://doi.org/10.1023/A:1004041530105

[56] Wotton RS.

[57] S.H. Abdel-Aziem, H.A.M. Abd El-Kader, F.M. Ibrahim, H.A. Sharaf, El A.I. makawy, Evaluation of the alleviative role of Chlorella vulgaris and Spirulina platensis extract against ovarian dysfunctions induced by monosodium glutamate in mice, J. Genet. Eng. Biotechnol. 16(2) (2018) 653–60.
https://doi.org/10.1016/j.jgeb.2018.05.001

[58] P. Libby, M. Nahrendorf, F.K. Swirski, Monocyte heterogeneity in cardiovascular disease, Semin. Immunopathol. 35(5) (2013) 553–62. https://doi.org/10.1007/s00281-013-0387-3

[59] J. Jeevanandam, A. Barhoum, Y.S. Chan, A. Dufresne, M.K. Danquah, Review on nanoparticles and nanostructured materials: History, sources, toxicity and regulations, Beilstein J. Nanotechnol. 9(1) (2018) 1050–74. https://doi.org/10.3762/bjnano.9.98

[60] S. Marzorati, L. Verotta, S.P. Trasatti, Green corrosion inhibitors from natural

sources and biomass wastes, Molecules. 24(1) (2019) 48.
https://doi.org/10.3390/molecules24010048

[61] S. Perumal, S. Muthumanickam, A. Elangovan, R. Karthik, K.K. Mothilal, Bauhinia tomentosa leaves extract as green corrosion inhibitor for mild steel in 1M HCl medium, J. Bio.-and Tribo.-Corros. 3(2) (2017) 13. https://doi.org/10.1007/s40735-017-0072-5

[62] H.A.R. Suleria, C. Barrow, Bioactive Compounds from Plant Origin: Extraction, Applications, and Potential Health Benefits, CRC Press, 2019.

[63] L.L. Shreir, Localised Corrosion, in: Corrosion, L.L. Shrier, R.A. Jarman, G.T. Burstein, v 1. Oxford, Butterworth-Heinemann; UK, 1994. https://doi.org/10.1016/B978-0-08-052351-4.50014-5

[64] R.W. Bosch, J. Hubrecht, W.F. Bogaerts, B.C. Syrett, Electrochemical frequency modulation: a new electrochemical technique for online corrosion monitoring. Corros. 57(1) (2001) 60–70. https://doi.org/10.5006/1.3290331

[65] A. Edreva, V. Velikova, T. Tsonev, S. Dagnon, A. Gürel, L. Aktaş, Stress-protective role of secondary metabolites: diversity of functions and mechanisms. Gen. Appl. Plant Physiol. 34(1–2) (2008) 67–78.

[66] H.U. Blaser, H.-J. Federsel, Asymmetric catalysis on industrial scale: challenges, approaches and solutions, John Wiley & Sons, 2011. https://doi.org/10.1002/9783527630639

[67] L. Vrsalovic, D. Sardelic, S. Gudic, M. Kliskic, Influence of green tea extract on corrosion of different metals in 0.5 mol dm^{-3} NaCl solution, J. Adv. Chem. 10(10) (2014). https://doi.org/10.24297/jac.v10i10.872

[68] X.C. Ma, X. Xue, A. González-Mejía, J. Garland, J. Cashdollar, Sustainable water systems for the city of tomorrow—A conceptual framework, Sustainability 7(9) (2015) 12071–12105. https://doi.org/10.3390/su70912071

[69] M.G. Lobo, E. Dorta,Utilization and Management of Horticultural Waste, in: Postharvest Technology of Perishable Horticultural Commodities. Elsevier, 2019. pp. 639–666. https://doi.org/10.1016/B978-0-12-813276-0.00019-5

[70] A. Al-Karaghouli, L.L. Kazmerski, Energy consumption and water production cost of conventional and renewable-energy-powered desalination processes, Renew. Sustain. Energy Rev. 24 (2013) 43–56. https://doi.org/10.1016/j.rser.2012.12.064

[71] M.I.A Sagiroun, Concentration of some radionuclides in some popular sudanese medicinal plants, Theses, 2012.

[72] L. Siauciunaite-Gaubard, Exploration de nouvelles approches pour les études de RCPG au niveau moléculaire: application aux récepteurs de chimiokines, 2012.

[73] N. El Hamdani, R. Fdil, M. Tourabi, C. Jama, F. Bentiss, Alkaloids extract of Retama monosperma (L.) Boiss. seeds used as novel eco-friendly inhibitor for carbon steel corrosion in 1 M HCl solution, Electrochem. Surf. Studies. Appl. Surf. Sci. 357 (2015) 1294–305. https://doi.org/10.1016/j.apsusc.2015.09.159

Theory and Applications of Green Corrosion Inhibitors
Materials Research Foundations **86** (2021) 91-126

Materials Research Forum LLC
https://doi.org/10.21741/9781644901052-4

Chapter 4

Electrochemical Studies of Green Corrosion Inhibitors

S.K. Ujjain[1]*, P. Ahuja[1], R. Kanojia[2]

[1]Research Initiative for Supra-Materials, Shinshu University, 4-17-1 Wakasato, Nagano-City 380-8553, Japan

[2]Department of Chemistry, Shivaji College, University of Delhi, India

* drsanjeevkujjain@gmail.com

Abstract

Corrosion of macro to micro-structures has been one of the major causes of structural failure in modern era. Their early detection could assist in limiting their repairs and reducing the associated cost as well. This chapter focusses on the state-of-the-art and development made with green corrosion inhibitors for preventing corrosion. It mainly includes the most recent progress in electrochemical corrosion monitoring techniques for various green inhibitors namely polarization technique, electrochemical impedance spectroscopy, electrochemical noise measurement and electrochemical quartz crystal microbalance. Finally, we conclude with the current progress, limitations and remedies in the recent trends along with advancement of green corrosion inhibitors for corrosion prevention.

Keywords

Electrochemical Techniques, Corrosion, Green Corrosion Inhibitor, Polarization Technique, Electrochemical Impedance Spectroscopy, Electrochemical Noise, Electrochemical Quartz Crystal Microbalance

Contents

1.   Introduction

The corrosion is related to the deterioration of the materials properties during its interaction with the environment. It is actually the result of the chemical reaction between a metal and its surrounding. It takes place on the metal surface owing to the electrochemical reactions in the atmosphere, which continuously caused the uniform annihilation of that metal's surface [1].

When describing corrosion, metallic corrosion is the most prominent to consider. Also, non-metallic materials like plastics, ceramics, etc. are susceptible to corrosion during their exposure to diverse corrosive surroundings [2]. Corrosion can also occur in engineering structures not only in macro scale dimensions but also in microelectronics. Corrosion in microelectronics is equally harmful as smaller amounts of dissolution in integrated circuits lead to the failure of larger systems. Even high-performance structural components may encounter corrosion (any form) regardless of the kind of material. Hence, it is imperative to analyze the kind of corrosion to which a particular material is susceptible. Generally, there are five types of corrosion which can occur on metallic surfaces say, general, stress cracking, localized, galvanic, and caustic agent corrosion. General corrosion occurs due to rust, specifically in steel. When metal come across with humid environment or directly in contact with water, the surface gets oxidized with the formation of a thin layer of rust. Stress corrosion, most damaging one, takes place under extreme tensile stress. As a result, metal component may develop cracks and become target for further corrosion. This form of corrosion mostly results from the stress caused by welding and thermal treatment. Localized corrosion occurs at particular region of a component owing to specific corrosion causing stresses. Galvanic corrosion occurs in dissimilar or similar metals with different electrochemical charges when they are connected via conducting path so that electron can move from anodic to cathodic part. Caustic agent corrosion results from the interaction with impure gas, liquids, or solids. Indeed, impure gases do not cause harm to metals in dry form but when they encounter humid surroundings, they resulted in harmful corrosive conditions [3,4].

In addition, storage of nuclear waste for prolonged time period may involve corrosion in terms of high level radioactive decay resulting in massive economic loss. This consequently made it the most challenging issue for developed and developing countries as well. Indeed, their impact goes afar from the metals and encompasses to energy and water. Not only direct loss such as replacing the corroded components of machinery say pipelines, metal roofing; quite large indirect losses are also involved, which are difficult to assess. Indirect losses mainly involve unexpected halt of machinery, loss of water or gas via corroded pipeline, efficiency loss in the energy conversion systems, contamination of the products in metal piping and different special requirements for designing equipment which can bear the applied stress with extended lifetime [5,6].

Corrosion cost determination using different methods developed by different scientists all over the world, mainly Uhlig method designed by Herbert H. Uhlig in 1949 [7], Hoar method developed by Hoar in 1971 [8], and economic input/output model invented by National Bureau of Standards (NBS) in 1978 [9] have been employed by nations including Germany, China, the United Kingdom, India, Finland, Japan, Australia, Kuwait, and the United States [10]. The resulted corrosion cost indeed ranged from 1–5% of gross national product (GNP) of particular nation depending on adopted methodology and the essentials of particular country [11]. This typically estimated cost do not included the environmental consequences or human safety, which can be very dangerous. This destruction has been unnoticed as the indication of proof got vanished during the catastrophic situations. As such, the collapse of the Silver Bridge is among the major hazardous corrosion accidents [12]. Also, the bridge linking Point Pleasant, West Virginia and Kanauga, Ohio unexpectedly crumpled, claiming many lives. The Bhopal accident happened in India in 1984 is among worst accidents with regard to human live injuries. Hence, the corrosion engineers should provide management techniques regarding the design and construction for material selection and environment alteration to prevent the major economic and ecological losses [13].

In order to have better understanding about the occurrence and prevention of corrosion, we should have in depth knowledge of why corrosion occurs and the driving force behind it. This further require thorough analysis that whether they are chemical or electrochemical. Metals have a strong tendency to revert to its lower energy state, i.e. metal oxides, which is responsible for corrosion to occur. The extent to which a metal undergo corrosion not only depends upon its electrode potential but also on the pH of the solution, which is well summarized in Pourbaix diagram [14]. Chemical reactions involve addition or removal of elements from the chemical species with no change in its valence state, i.e., no oxidization or reduction of chemical species take place. On the contrary, electrochemical reactions beside addition or removal of elements, also involve change in the valence of at least one species. Indeed, a wide variety of manufacturing ingredients dissolves or deteriorates via electrochemical reactions. For these electrochemical reactions to occur for corrosion, four basic requirements should be fulfilled involving an anode (oxidation occurs), cathode (reduction), an electrolytic and electrical path between anode and cathode for ionic and electronic conduction.

As a consequence, corrosion control measures must satisfy one of these requirements such as the use of oxygen scavengers (affecting cathodic reaction rate) and also, the use of organic coatings (inhibiting the electrolytic path formation). So, assessing the corrosion procedure involves the continuous analysis of the effects on one of the four requirements. Corrosion monitoring is thus attaining statistics on the kinetics of corrosion-induced mutilation to the component and is usually related to inspection of the component state at any instant of time. In today's era, different industrial processing parameters say temperature, pH and flow rates are organized by feedback controllers and reliable sensors. They make it possible to accurately manage the corrosion process and hence product quality is improved. Regrettably, the corrosion monitoring process is still at the infancy stage in most of the countries. Several successful strategies include the use of corrosion inhibitors for preventing the loss caused by corrosion. Corrosion inhibitors are chemicals which can diminish the rate of corrosion by preventing the metal dissolution with minimal environmental impact. Depending on their chemical nature, inhibitors can be categorized into organic and inorganic inhibitors whilst now-a-days these corrosion inhibitors are recommended based on their ecological aptness. Also, related agencies in many countries have implemented various rules and regulations regarding their usage. They require ecofriendly and safe corrosion inhibitors, which are discussed in later section.

The more important concern is the parameters based on which these inhibitors has been added to the systems, which are determined from typical results obtained by the test coupons exposed in the controlled environment. This can be resolved with the improvement in corrosion monitoring procedures based on kinetics of electrochemical processes. This chapter is intended to provide a platform for the scientists concentrating on green inhibitors for corrosion and their electrochemical monitoring for corrosion inhibition.

2.   Corrosion inhibitors

Corrosion inhibitors are chemicals which can diminish the metal's dissolution rate when added in corrosive environment [15]. They can provide provisional and localized shielding during transportation and storage to prevent corrosion arises from accumulation of an aggressive medium [16]. They can act in different ways i.e. (i) by forming a protective film on the metal surface, (ii) by resulting in products which can act as passivators, and (iii) by precipitating different moieties that can annihilate or deactivate a corrosion causing component [17].

Based on the prevention by four basic requirements of corrosion (stated earlier), these inhibitors can be classified as anodic, cathodic, or mixed type [18]. Anodic inhibitors diminish reactions at anode due to formation of partially soluble deposits like oxides or hydroxides, and hence can also be called as passivation inhibitors. On the contrary, cathodic inhibitors reduced the cathodic reaction rate by forming a shield to counter hydrogen in acidic environments and oxygen in alkaline surroundings. Different from above, mixed inhibitors have impact on both, the cathodic and anodic reaction owing to defensive film on the surface of metal.

Based on their chemical nature, they can be categorized into organic and inorganic inhibitors [19]. The capability of organic inhibitors depends on its adsorption capacity, along with mechanical and chemical physiognomies of the adsorption layer, resulting in a particular milieu [20]. In addition, it should have polar functional moieties with Nitrogen (N), Oxygen (O) or Sulphur (S) atoms along with hydrophobic entity, which can resist the metal surface from corrosive moiety [21,22].

The inorganic inhibitors have inorganic active substance for their potent inhibitive strength. Many inorganic anions, such as chromates (CrO_4^{2-}), phosphate ($H_2PO_3^-$), molybdate (MoO_3^-) and nitrates (NO_2^-) can offer passive defense to the metal surfaces via incorporating in the oxide layer. Although they provide potential inhibition efficiency, yet they have been proven hazardous for human health in the long run [23]. Also, the environmental and ecological menaces are concomitant with the organic inhibitors. The adverse influences on mankind and the environment has discouraged the usage of these inhibitors [24], which prompted the related researchers towards investigating non-toxic or green corrosion inhibitors [25].

2.1   Green corrosion inhibitors

The usage of green corrosion inhibitors should follow the basic requirements as suggested by   Registration, Evaluation, Authorization and Restriction of Chemicals (REACH) and Paris Commission (PARCOM) which include their non-bio accumulativeness, biodegradability, and very low marine toxicity [26]. Consequently, corrosion scientists are motivated towards the inexpensive, readily available, environmental and ecological friendly and renewable green corrosion inhibitors. These green corrosion inhibitors can fall in following classes:

2.1.1 Natural products

A large number of chemical moieties are present in plant extracts particularly heterocyclic, which efficiently inhibited the corrosion [27-29]. Among them are eco-friendly curing agents obtained from modified vegetable oils, which offer highly anticorrosive characteristics. Abdel-Gaber et al. have investigated plants extract corrosion inhibition efficiencies against steel in H_2SO_4 medium [30]. They demonstrated highest efficiency ~88% for black cumin with 1.40 g/l concentration. Also, ethanolic extract of Ricinus communis leaves were studied for mild steel (MS) and possess 85% efficiency with very low inhibitor concentration of 300 ppm [27,28]. Kliskic et al. inspected the aqueous extract of Rosmarinus officinalis against aluminum in HCl solution [31].

In addition, plant derivatives in the form of oils are also used to formulate coatings for metal surface for preventing corrosion. [32-35]. These inhibitors mitigate metallic corrosion by means of adsorptive layer of ions and molecules on the surface of metal, thereby forming a defense against the corrosion. Likewise, Jojoba oil activity was explored against iron and offered maximum efficiency ~100% at 0.515 g/l inhibitor concentration [36,37]. Furthermore, Rosemary oil also inhibited the corrosion of mild steel in H_2PO_4 with 73% efficiency [38,39].

Additionally, drugs have also been utilized as green corrosion inhibitors being non-toxic, inexpensive, and eco-friendly. Also, most of the existing drugs are obtained from natural source [40] and the existence of heteroatoms and heterocycles along with benzene impart corrosion inhibition activity to them. In recent times, Ali et al. examined the inhibition property of Candesartan drug on carbon steel (CS) in acidic medium. It showed high efficiency with 300 ppm concentration of the drug [41]. An inflammatory drug ketosulfone has been explored towards corrosion inhibition properties in acidic medium against mild steel (MS). Remarkable efficiency has been reported 96.6% at 200 ppm concentration [42]. Astonishingly, an expired atorvastatin drug also possesses the capability of 99.1% at 150 ppm inhibitor against mild steel in acidic medium [43].

2.1.2 Amino acids

Amino acids with carboxyl (-COOH) and amino (-NH$_2$) moieties at same carbon atom possess significant anti-corrosive effects. The existence of heteroatoms, such as N, O, and S along with conjugated π-electrons in the system prompted their potential corrosion inhibition ability [44,45]. El-Sayed et al. further inspected the anti-corrosive activity of few amino acids towards carbon steel in aerated chloride solutions. Presence of different moieties say hydroxyl (OH), mercapto (SH), or phenyl help them undergo better adsorption with high corrosion mitigation [46]. Mobin et al. explored the mitigation activity of amino acid L-cysteine (CYS) with SH group towards MS at 30°C, resulting in high efficiency 85.6% at 500 ppm [47].

2.1.3 Rare earth metal compounds

The pioneering work on applying lanthanides for protection of aluminum alloys was scrutinized by many researchers [48,49]. Different concentration of cerium trichloride (CeCl$_3$) has been successfully utilized in sodium chloride (NaCl) solution [50] and displayed a steep decrement in corrosion rate at 100 ppm whereas higher concentration produces stabilizing effect in rate of corrosion. Also, Liu explored the corrosion protection of Al–Zn–Mg alloy in NaCl solutions with CeCl$_3$ at different concentration and pH [51]. Likewise, Cerium nitrate (Ce(NO$_3$)$_3$) and Cerium sulfate (Ce$_2$(SO$_4$)$_3$) have also been investigated, where the inhibition activity of cerium is screened by nitrates and sulphates anions [52]. Researchers demonstrated that these rare earth salts form an oxide film at the metal surface, preventing oxygen or electron diffusion for the reduction reaction. Thereby, they can reduce the corrosion rate and successfully mitigate corrosion [53]. Not only cerium but to a lesser extent, other salts of Rare Earth elements such as Yttrium chloride (YCl$_3$), Lanthanum chloride (LaCl$_3$), Praseodymium chloride (PrCl$_3$) and Neodymium chloride (NdCl$_3$) have also been used to provide high degree of corrosion protection [54-57] by improving the adhering ability of oxide films on different metallic alloys.

Theory and Applications of Green Corrosion Inhibitors   Materials Research Forum LLC
Materials Research Foundations 86 (2021) 91-126   https://doi.org/10.21741/9781644901052-4

2.1.4 Recently used green inhibitors

Earlier reported natural products have been widely utilized as green corrosion inhibitors but they are associated with tedious, time-consuming and expensive extraction and purification processes [58]. Therefore, there is urgent need to develop green inhibitors with proper design consideration by using efficient starting materials or single step reactions. Regarding this, utilizing surfactants and ionic liquid-based inhibitors can provide a new strategic step. Surfactant based corrosion inhibitors also have potential to act as green corrosion inhibitors owing to their high efficiency, economic and non-toxicity. Surfactant molecules have a polar "head" and a non-polar "tail" with hydrophilic and hydrophobic nature, respectively. In aqueous solutions, they adsorb via either chemisorption or physisorption for inhibiting corrosion [59-62]. For instance, non-ionic dithiol-based surfactants, 2-mercaptoacetic acid and polyethylene glycol have been used to formulate di-mercaptoethanoate polyethylene glycol. It has been shown that corrosion efficiency increases with the molecular weight [63]. Also, combining cetyl trimethyl ammonium bromide (CTAB) with indigo carmine considerably augmented the corrosion inhibition of CS to 98.5% in acidic solution [64].

Additionally, a novel Gemini surfactant (Ethane-1,2-diyl bis(N,N-dimethyl-N-hexadecyl ammonium acetoxy) dichloride have also been explored by Mobin et al. which considerably increases the corrosion inhibition activity to 98% for MS in HCl at 60 °C [65].

Lately, ionic liquids have been also attracted as efficient green corrosion inhibitors owing to their unique attributes say low volatility, lower vapor pressure, high thermal stability and no combustibility along with their low toxicity [66,67]. They generally have organic cations with organic or inorganic anions and exist in liquid phase at room temperature [68]. Interestingly, their properties can be tailored from their structural perspective by adopting extensive range of possible groupings between cations and anions [69,70]. Existing literature have established that imidazole, ammonium, pyridine and phosphonium have high corrosion inhibition activity for protecting metal in destructive/corrosive environment [71]. Excitingly, quaternary ammonium based ionic liquids also possess high anticorrosive activity [72-74] which predominantly arises from the chemical construction of the cations. They involve physical adsorption on the metallic substrate as their inhibition mechanism.

3.   Characterization techniques

The mechanism of corrosion in aqueous phase actually involves electrochemistry. Hence most of the characterization techniques for corrosion monitoring involve electrochemical perspective. During the monitoring of corrosion via electrochemical experimentations, samples are polarized to record different measurements related to corrosion, within minutes or hours. This electrochemical corrosion reaction necessitates four elements i.e. anode, cathode, metallic and electrolytic conductor. Anode is specifically bound for oxidation where metallic ion departs from the surface and goes into the solution and this process is corrosion. The resulting metallic ions or the reactive species, then migrate to cathode by conducting electrolyte. In real-time measurements, anodic and cathodic sites can be presented on the same model and can also act as metallic conductor. So, when such metal or model is dipped in electrolyte, a potential is generated, which consequently results in corrosion to occur. The corresponding potential is called corrosion or mixed potential, which measures the corrosion tendency of particular sample (metal or alloy). This corrosion potential is also measured in reference with other electrode (reference electrode).

In order to study the kinetics of these corrosive electrochemical reactions at metal-solution interface efficiently, we should use the testing sample of interest as an electrode of electrochemical cell. These electrochemical cells can be classified as driving (power producer) or driven (power consumer) depending on their function [75]. However, corrosion systems are driving systems which are short-circuited. The anodic and cathodic sites are present on the same metal surface where corresponding anodic and cathodic reactions occur. For studying corrosion kinetics in the electrochemical systems with working, reference and counter electrode, a potentiostat is used. Generally, corrosion kinetics of working electrode can be investigated by controlling its potential in regard with reference electrode at a constant value and correspondingly, recording the reaction rates (current). If the voltage difference between reference and working electrode is different from the input voltage, output voltage will be non-zero until the voltage difference between working and reference electrode becomes zero. This allow the current to pass through counter electrode, solution and working electrode interface, whereas no current passes through reference electrode, maintain it at equilibrium to serve as potential electrode. If the current changes owing to alterations at working electrode surface say during formation of a film (to maintain input voltage), corresponding changes in the output voltage can also be observed instantaneously. Likewise, changing the polarity and magnitude of the input voltage, the kinetic behavior of an interface with regard to current-voltage (I-E) correlation can be examined over a range of potential. In addition, the consequences of external variables such as temperature, flow rate of solution and its composition can also be inspected under constant electrochemical conditions.

Potential of the system provides a signal for the corrosive propensity of metals. In order to know the rate of corrosion reaction, we should have information of the electron transfer rate between electrode and electrolyte since corrosion rate is proportional to it. Also, redox potential of a system is defined as the potential when a metal say Zinc (Zn) attains equilibrium with the solution of its own ions at unity concentration whereas redox potential at other concentrations can be calculated by using Nernst Equation [76]. The rate at which electrons exchanged under this condition is exchange current density. On the contrary, if Zn is immersed in H^+ ions solution, the potential stabilizes at corrosion potential (E_{corr}). At E_{corr}, the rate at which Zn dissolution takes place is approximately same as the rate at which hydrogen evolution takes place. The current at E_{corr} is represented by I_{corr} for simplicity, which is a measure of the rate at which Zn will corrode when it is dipped in solution of H^+ ions.

To establish I-E relationship, there exist several electrochemical methods in which electrochemical excitation are applied to an electrode and corresponding response will be monitored. As a result of this excitation, the electrode possesses different potential from corrosion potential and said to be polarized. So, they are correspondingly called polarization methods for corrosion monitoring. Most of these experiments are preferred by regulating the potential (potentiostatic or potentiodynamic) instead of current.

3.1 Polarization methods

3.1.1 Linear polarization resistance method

In this, potential from 10 to 30 mV is provided to the electrode and the resultant change in current is monitored. As very small perturbation potential is applied, this method does not provide obstructions in corrosion reactions. The ratio of current to potential is polarization resistance (R_P) [77] and can be calculated from the slope of potential-current plot at corrosion potential as shown in Fig. 1 and Eq. 1:

$$R_P = \left(\frac{\Delta E}{\Delta I}\right)_{E_{corr}} \tag{1}$$

Also, this polarization resistance is associated to the corrosion current, i_{corr} by Stern-Geary equation given by [78]:

$$\frac{\Delta E}{\Delta I_{app}} = \frac{\beta_a \beta_c}{2.303 I_{corr}(\beta_a + \beta_c)} \tag{2}$$

Where β_a and β_c are Tafel slopes observed from anodic and cathodic reactions, respectively. ΔE is applied potential change, Δi_{app} is corresponding change in current density and i_{corr} is corrosion current density at corrosion or mixed potential.

Materials Research Forum LLC
https://doi.org/10.21741/9781644901052-4

In real life applications, commercially available probes are placed in the container in which corrosion reaction is taking place to determine the corrosion rate. A power supply polarizes the electrode and corrosion rate is monitored by recording the resulting change in current. These probes can then be connected with computerized data acquisition module and an alarming system can be developed for vigilance by alerting the operators about any discrepancy of large corrosion rates [79,80].

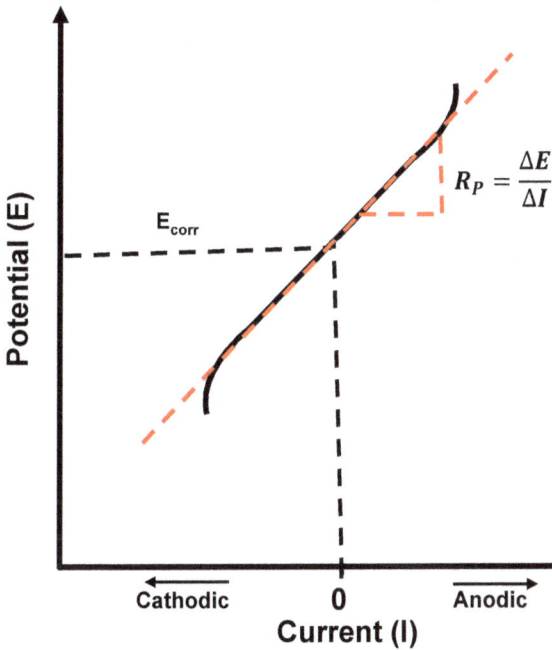

Figure 1. Schematic illustration of linear polarization test by Potential-current plot (R_P = Polarization resistance, E_{corr} = Corrosion potential).

Furthermore, it is interesting to note that current-potential curve may not always be linear in the region of corrosion potential. Also, Anodic, or cathodic sections may not be symmetrical if β_a and β_c are not equal. Then, they are determined by Tafel extrapolation method.

Linear polarization method was used to estimate the corrosion rate of the zinc with Mansoa alliacea plant extract as corrosion inhibitor. Corrosimetry (the time profile of R_p) was utilized to study the corrosion inhibition activity from the following formula [81]:

$$Inhibition\ Efficiency\ (\%) = \left[\frac{R_P - R_P^o}{R_P}\right] \times 100 \qquad (3)$$

The results obtained from R_p versus time (490 Ω.cm^2) are in good agreement with those obtained from polarization curves (517 Ω.cm^2). The R_p values have been used to calculate the inhibition.

3.1.2 Potentiodynamic-galvanodynamic polarization

Three electrode system is used in this technique to polarize the working electrode. The potential (potentiodynamic) or current (galvanodynamic) is then varied corresponding change in current or potential is monitored. Here, the applied potential perturbation is of few 100 mV unlike that in linear polarization resistance method. Mostly potentiodymanic measurements are preferred over galvanodynamic and are more beneficial to evaluate anodic and cathodic Tafel slopes from extrapolation method.

For that, Tafel established a linear relation between log I and E by polarizing the electrode to adequately large potentials in both anodic and cathodic directions which is given by the following equation [82]:

$$I = I_{corr} \left[\exp\left\{\frac{2.303(E - E_{corr})}{\beta_a}\right\} - \exp\left\{-\frac{2.303(E - E_{corr})}{\beta_c}\right\}\right] \qquad (4)$$

Where E and I are applied potential and current, E_{corr} and I_{corr} is potential and current at corrosion potential, β_a and β_c are Tafel slopes in anodic and cathodic regions from E-log I plot. Metals which are prone to corrosion, in general show more or less similar Tafel behavior as shown in Fig. 2 when polarized [83].

For simplifying the above relation, the difference between E and E_{corr} is defined by overpotential, η. At larger values of overpotential, in anodic direction, η_a is displayed by the following equation from Tafel equation [84]:

$$\eta_a = \beta_a \log\frac{I}{I_{corr}} \qquad (5)$$

Similarly, in cathodic direction, overpotential is represented by:

$$\eta_c = -\beta_c \log\frac{I}{I_{corr}} \tag{6}$$

In the Tafel regions, I_{corr} is finally observed by extrapolating either anodic or cathodic slopes to E_{corr} as shown in Fig. 2.

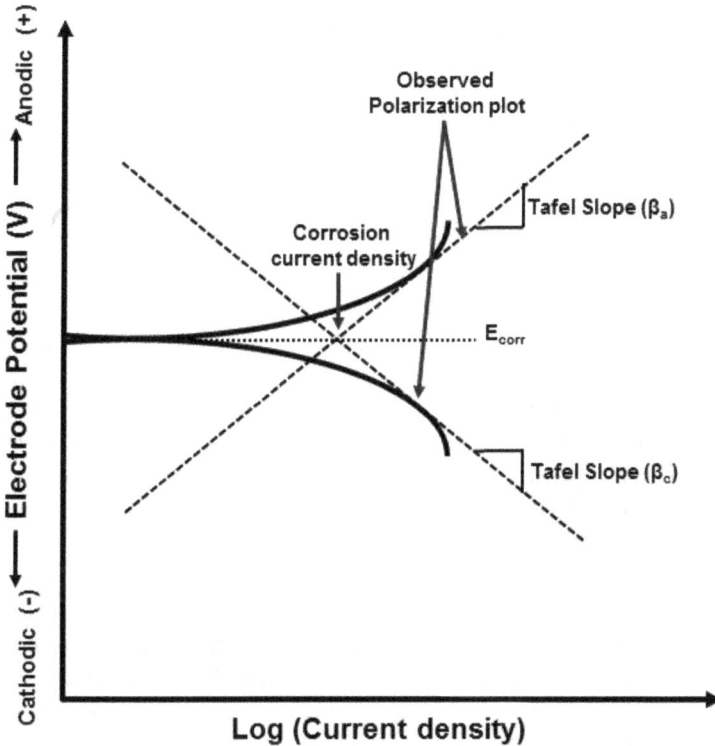

Figure 2.   Tafel extrapolation method for the cathodic and anodic polarization curves for determination of corrosion current density and Tafel slope.

In this method, stabilized value of E_{corr} is firstly measured in course of time followed by potential step or potentiodynamic measurements. In potential step method, incremental potential of ± 25 mV from 25 to 100 mV is applied after every five minutes and corresponding current at the end of each step is noted. Whereas in the potentiodynamic method the potential is scanned at a constant rate. In both cases, the experiment start from corrosion potential and conducted with cathodic and anodic polarization of the electrode. Then, both the regions are plotted in semi-log plot of E vs log I in the Tafel plot. From the resulted Tafel plot, we can determine anodic and cathodic Tafel slopes and I_{corr} [85].

As this method involves the application of large overpotential to the electrode, it is regarded as damaging approach, particularly during anodic polarization where surface may get permanently damaged. Hence, it is not recommended as a corrosion monitoring technique in practical applications. Also, the rate of applied signal may result in the delayed measured signal behind the scheduled or true steady-state signals or data. This consequently results in disagreement in practical and ideal time required for the whole scan and also the lagging extent in observed and ideal response. Hence, this technique proved unsuitable for real time corrosion monitoring where continuously varying environmental conditions exist.

3.1.3 Cyclic potentiodynamic polarization

It is generally used to measure the propensity of metals or alloys to pitting corrosion, which is a localized corrosion by which cavities or "holes" can be produced in the material [86]. The potential is initially scanned in the noble direction (positive) until a specified anodic current is achieved. Then, the scan direction is reversed (negative) till the initial potential is reached (hysteresis loop closes). The potential is observed at a fixed scan rate where hysteresis loop is completed and the more electropositive is the potential, less prone is the material to localized corrosion. If no hysteresis is observed, there is no existence of localized corrosion instead there occur uniform corrosion [87].

Some other parameters can also be determined from this method say primary passivation potential, positive to which inactive surface layers are formulated. Also, breakdown potential can be observed where inert surface layer is demolished and transpassive region has originated. On the contrary, protection potential is also imperative as at this potential, inert layers are stable and protected, which can also be calculated by utilizing empirical observations [88].

Potentiodynamic polarization analysis of aluminum in biodiesel is carried out with and without Rosemary extract at 298 K [89]. It was observed that the Rosemary extract showed decreased anodic and cathodic current densities owing to the adsorption of the extract's organic components at the active sites of the electrode surface. Also, in the presence of Rosemary extract, E_{corr} shifted anodically with respect to the blank indicating predominant anodic effectiveness of the mixed-type Rosemary extract inhibitor [90].

3.1.4 Cyclic galvano-staircase polarization

This method is more applicable to determine the proneness of a material to localized corrosion i.e. by protection potential. In this, stepwise current signal is applied to forward and backward direction with a particular step height and duration and the corresponding potential variations are monitored at each step. Then these step up and step-down data points are extrapolated at zero current to acquire breakdown and protection potential respectively. This method only provides qualitative information regarding pit initiation but not on pit propagation.

Polarization study of aluminum in basic medium was carried out, without and with different concentrations of Piper longum extract as green corrosion inhibitor [91]. The corresponding inhibition efficiency was calculated from I_{corr} by using following equation [92]:

$$Inhibition\ Efficiency\ (\%) = \left[\frac{I_{corr} - I_{corr(i)}}{I_{corr}}\right] \times 100 \tag{7}$$

Where I_{corr} and $I_{corr(i)}$ represent the corrosion current density during absence and presence of corrosion inhibitor, respectively.

Here, β_c decreases with increasing inhibitor concentration, indicating the formation of P. Longum extract film on the metal surface and corresponding hindrance to the alkali attack on the aluminium electrode. On the contrary, anodic Tafel slope showed shift which can be accredited to the alteration in anodic dissolution process owing to the adsorption of inhibitor molecules at active sites. The cathodic and anodic curves obtained from the working electrode in presence of P. Longum extract showed shift in the direction of current reduction, as observed from polarization results which increased the inhibition efficiency (IE%) with maximum extract concentration of 89% at 400 mgL^{-1} inhibitor concentration.

3.1.5 Conversion of I_{corr} (from polarization methods) to corrosion rates

Corrosion current (I_{corr}) measured from the above said methods is then converted to corrosion rate in order to determine the efficiency of particular inhibitor. In order to realize that effectively, it was supposed that the current distribution is non-deviating in the concerned area (A). Keeping that in mind, current density (i_{corr}) is calculated as follows:

$$i_{corr} = \frac{I_{corr}}{A} \tag{8}$$

Based on Faraday's laws [93], the corrosion rate or mass loss rate equation [94-96] is as follows which can be used to calculate the rate associated with corrosion:

$$Corrosion\ Rate = K_1 \left(\frac{\rho_{corr}}{\rho}\right)\left(\frac{W}{n}\right) \tag{9}$$

Where corrosion rate is expressed in mm/y, K_1 is 3.27×10^{-3} (mm g/μA cm y), ρ is the density (g/cm^3) and W/n is equivalent weight (It is the mass of pure element which is oxidize by one Faraday of electric charge). Here, W represents element's atomic weight and n is number of electrons essential for oxidizing an atom of element. And mass loss rate is given by:

$$Mass\ loss\ Rate = K_2 i_{corr} \left(\frac{W}{n}\right) \tag{10}$$

Where mass loss rate is expressed in g/m^2, K_2 is 8.9×10^{-3} gcm^2/μAm^{2}d, during the corrosion process.

3.1.6 Limitations associated with polarization methods

There are several disadvantages of utilizing these polarization techniques. Now as we all know that in these polarization techniques, we apply current or potential and correspondingly monitor potential or current respectively. With all these calculations, polarization resistance is associated to corrosion. But the measured resistance is composed of two constituents i.e. solution resistance (R_s) and polarization resistance (R_p). If R_s is high, then corrosion rate will be underestimated and vice versa. This error is generally considerable in systems with high corrosion rate and low conductivity.

There can also be significant error related with the higher scan rate owing to which corrosion rate is overestimated. The corroding element may be contemplated as a capacitor and when potential is applied, electric charge is build up as capacitance. Therefore, capacitance current component becomes higher than corrosion current or Faraday current at higher scan rates and consequently corrosion rate is over evaluated. It can be minimized by recording the polarization experiments at lower scan rates (~0.6 V/h).

In addition, the presence of redox species in the electrolyte interferes with the corrosion reaction, where they preferably undergo reaction. Under these circumstances, measurements of corrosion rates are not at all reliable. Also, corrosion potential should be scrutinized for adequately larger time to obtain stabilized value for using efficient and reliable electrochemical techniques. Moreover, localized corrosion may bring variation in the electrode geometry or related area resulting in variable polarization response, which is impossible to quantify. Importantly, corrosion reaction of the system should be charge-controlled and not diffusion-controlled as under diffusion-controlled corrosion, polarization methods will not provide accurate results. During high charge transfer rate, reduction species is rapidly exhausted at cathode, where the rate becomes diffusion dominated.

3.2 Electrochemical impedance spectroscopy (EIS)

In order to compensate the errors or nullify the limitations of polarization methods, it is necessary to examine the system considering general aspect of dynamic response to small signal excitations. This can be successfully achieved by Electrochemical Impedance Spectroscopy (EIS) measurements and directly used for evaluating corrosion rates. The principle of this technique involves the application of alternating signal (5-20 mV) to the working electrode [97]. Corresponding measurements of amplitude and phase shift in current and voltage component were made at various frequency in the spectral range (1 mHz-1 MHz) [98-100]. Notably, the initial excitation is sinusoidal potential (ΔE) which is executed at steady state and the corresponding response is sinusoidal current (ΔI) along with a phase shift or phase difference (φ) from the applied signal. Hence, the impedance is measured with ΔE and ΔI by using Ohm's law. It is indeed a non-destructive technique as small amplitude signals are applied which will not significantly affect corrosion potential.

The observations in the frequency domain from this technique is reflected as an interface which can be considered as a series or parallel combination in equivalent circuit involving passive electrical circuit elements say inductance, resistance, and capacitance. This equivalent circuit can be fitted with EIS results using Nova 1.7 and Zview software, allowing the discussion of the corrosion behavior in terms of different damage indices [101]. However, interpreting the results sometimes proved difficult as equivalent circuit can change depending on working electrode conditions [102]. More importantly, it can provide the information regarding electrochemical mechanism i.e. corrosion is whether activation, concentration, or diffusion dependent.

3.2.1 Interpretation of results (Nyquist & Bode plots)

EIS measurements can generally be considered by two plots, Nyquist and Bode plot. In Nyquist plot, real (Z') and imaginary (Z'') impedance are plotted at different frequencies and correlated with equivalent diagram as shown in Fig. 3. This Nyquist plot is also called as Argand representation or Cole-Cole plot. It has complex plane of Cartesian coordinates where X-axis is real part with resistive contribution and Y-axis shows imaginary part with inductive or capacitive contribution. Nyquist diagram generally has a semicircular arc in high frequency region and followed by spike in low frequency region, depending on the working electrode situation [98-100]. The diameter of the semicircular arc when extrapolated to X-axis represents the charge transfer resistance (R_{CT}), which is generally similar to R_P [103]. It can be directly related to the corrosion rate as larger diameter of semicircle has large R_{CT} or R_P and correspondingly lower corrosion rate [104]. The bode diagram has two quantities of impedance on Y-axis i.e. logarithm of impedance, modulus ($\log |Z|$) or phase angle (φ) with logarithm of frequency ($\log f$) on X-axis. It is generally used to calculate the similar parameters and, double layer capacitance (C_{dl}) more precisely [105,106].

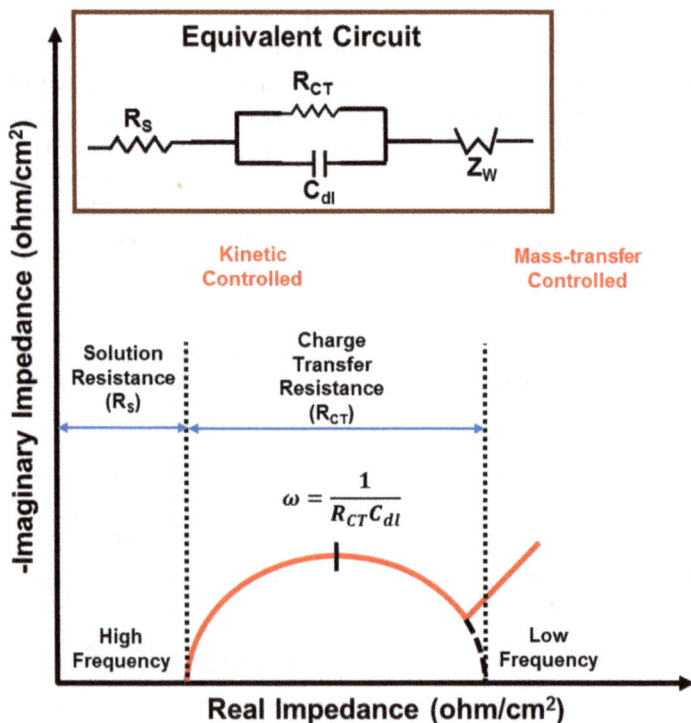

Figure 3.   Electrochemical Impedance Spectroscopy Nyquist plot and Equivalent circuit.

The corrosion of low CS in 1 M HCl in the presence of green inhibitor (Schinopsis lorentzii extract) was inspected by EIS at room temperature after 30 min of immersion [107]. Nyquist plot was significantly changed on adding the inhibitor and corresponding value of charge transfer resistance was obtained after fitting equivalent circuit. R_{CT} values showed increase in inhibition efficiency with good agreement with polarization measurements. It can be concluded that the corrosion rate depends on the chemical nature of the electrolyte rather than the applied technique [108]. In addition, MS corrosion has been studied in presence of Zenthoxylum alatum plant extract by impedance spectroscopy [109]. It was demonstrated from Nyquist plot that Capacitance (C_{dl}) values decreases and R_{CT} increases on increasing the plant extract concentration indicating decrease of charges stored in the double layer owing to the development of chemisorbed extract layer which correspondingly increases the corrosion inhibition.

Interestingly, the effect of temperature on corrosion inhibition efficiency of Oxandra asbeckii plant extract was observed by studying EIS between 25–55 °C [110]. Examination results revealed that the temperature rise lead to decrement in R_{CT}. It is due to increment in metal dissolution rate and the adsorption/desorption equilibrium shift towards the inhibitor's desorption and hence surface coverage decreased. Also, they have calculated activation energy and enthalpy from impedance data at different temperature, which showed good agreement with Tafel measurements.

3.2.2 Equivalent circuits

Corrosion process involves the simultaneous occurrence of many physical processes, corresponding to which equivalent circuit can be made with various circuit elements. However, depending on the particular process, the connection between the circuit elements may vary. The pioneers of EIS for corrosion monitoring proposed different models depending upon the particular situation [111,112]. If the model shows good agreement between the experimental data and the Nyquist diagrams, the data can be elucidated from the software generating the equivalent circuit.

Regardless of increasing advancement in the interpretation of the technique, sometimes it is difficult to explain the behavior demonstrated in the plots and equivalent circuit. Challenges often arise from low frequency region branches, depressed or displaced semicircle and high frequency effects [113,114]. In general, impedance response in equivalent circuits can be expressed in terms of three frequency bands related to different processes:

High frequency (MHz-kHz): Response is dominated by the electrolyte i.e. solution and charge transfer resistance can be analyzed [115,116] and the capacitance is in the order of pF/cm^2. Medium Frequency (kHz-Hz): Faradaic reactions are responsible for the impedance response in this region and capacitances follow in the order of $\mu F/cm^2$. Sometimes the response of high frequency region extends to this and flattening of the semicircle can be obtained. Low Frequency (Hz-mHz): Here the response arises owing to the redox reaction occurring in the system and typical capacitance falls in mF/cm^2 range.

Based on these, Christensen et al. projected an association between equivalent circuit and corrosion phenomenon, which is widely accepted in order to monitor the corrosion process [117,118].

3.3 Electrochemical Noise (EN) measurements

As a result of corrosion such as pit origination, electrode potential or current of the system fluctuates with time and these stochastic fluctuations are known as electrochemical noise [119]. This technique involves the measurements of these expected variations or noise of current or potential under applied potential or current respectively [120,121]. These noises were then monitored in potentiostatically polarized conditions and in freely corroding conditions [122,123]. Also, no additional corrosion effects are induced during measurement of the noise at open circuit. Examination of these fluctuations over a period and even after spectral decomposition provides indirect average corrosion rate.

These measured values then transformed to frequency domain, followed by estimation of power-density spectrum (PSD) values. PSD are plotted as base 10 logarithm in decibels vs log of frequency and the corresponding results were compared with I vs E measurements. The corrosion current density (i_{corr}) for comparison was estimated from Tafel extrapolation and correlation coefficient was obtained. Then the general corrosion rate was expressed by corrosion current density with which electrochemical noise shows good correlation [124]. So, it is always desirable to use this procedure in unification with other methods. More importantly, different types of corrosion can also be monitored as different shapes of electrochemical voltage and current signals owing to different type of corrosion resulted in different reflections in PSD.

An eco-friendly corrosion inhibitor from Thymus vulgaris plant extract was explored for their corrosion inhibition activity by using EN measurements [125]. The obtained data is also evaluated in frequency domain by figuring the PSD via Fast Fourier transformation (FFT) algorithm. Then, the bode and PSD diagrams are compared and obtained results displayed lower noise magnitude in PSD diagrams indicating useful information can be obtained particularly in lower frequency region. Furthermore, Mehdipour et al. and Hermoso-Diaz et al. studied the inhibition action of Aloe Vera leaf extract and Salvia hispanica on stainless steel (SS) and CS, respectively. Authors have discussed the use of the Stochastic fluctuations of potential or current occurring at the electrode/solution interface to derive the noise resistance (R_n) and IE. They observed the time records of electrochemical current noise associated with SS electrode dipped in 1 M H_2SO_4 with different concentration of Aloe Vera leaf extract. The current rise and drop were associated with rapture and recovery of passive inhibitor layer over the surface of SS. They observed that the amplitude of current transients show decrement with increasing concentration of Aloe Vera extract and attained a minimum difference between current fluctuation. Their EN data demonstrated close agreement with the DC polarization and EIS measurement results [126,127]. In addition, Shahidi et al. used wavelet transform (WT) mathematical tool for EN analysis and obtained the inhibition efficiency of Vanillin on stainless steel surface in 0.5M H_2SO_4 solution. The IE values derived as such exhibit close agreement to the IE derived from the potentiodynamic polarization and EIS measurements [128]. Analysis of such EN data have also been used for affirming the corrosion inhibition of herbal compounds or heterocyclic organic compounds for copper exposed to chloride and acid media, respectively [129,130].

3.4   Electrochemical Quartz Crystal Microbalance (EQCM)

It has been applied to the study of the corrosion of different metals [131,132], owing to its high sensitivity to mass changes [133]. In 1880, Jacques and Pierre Curie realized that when mechanical stress was applied to Quartz crystal surface, an electrical potential was generated whose magnitude was proportional to the stress [134]. This is called Piezoelectric effect. On the contrary, the application of potential to this crystal resulted in mechanical strain, which is used as a principle in EQCM technique [135]. This is highly useful for in situ findings and monitoring of the material's surface at the microscopic level. The typical configuration includes the placement of quartz crystal in between the 'O' shaped rings in which one side is in contact with the solution of electrochemical system. The electrode in contact with solution acts as QCM oscillator circuit and as working electrode in electrochemical measurements.

Sauerbrey described the variation in resonance frequency of a quartz crystal when a foreign mass is deposited on the top of the oscillator according to the following equation [136]:

$$\Delta f = -\frac{2f_0^2}{z_q}\Delta m \qquad (11)$$

where Δf shows quartz crystal resonance frequency shift, Δm is rigid mass density deposited on one side of the resonator, Z_q is the quartz mechanical impedance and f_0 is the unloaded quartz resonance frequency. So, by measuring the shift in the resonance frequency of the quartz crystal, we can obtain in-situ information of the deposited mass.

To apply Sauerbrey work, certain conditions should be kept in mind regarding the deposited film that it is indeed smooth and rigid [136]. Additionally, it is very important to separate the reason for frequency shift either from the viscoelasticity of the solution or from the mass change of quartz resonator. Also, if the shift in the resonance frequency is higher than the shift in damping, Sauerbrey equation can be successfully applied. Else, some rectification terms must be considered and added [137-139].

In addition, EQCM, when combined with other surface analysis techniques and EIS made it possible to study the mechanisms of various corrosion reactions, such as growth and dissolution of passive oxide films of metals or alloys in corrosive environments [140]. In a study, Kern et al. applied rotating electrochemical quartz crystal microbalance (rEQCM) having exchangeable electrode holders mounted on a rotating shaft to investigate the formation of protective coatings of aromatic carboxylic acid inhibitors for MS. Under this experiment condition, the change in frequency (Δf) is constituted of two contributions, i.e. the frequency shift of species adsorbed at the electrode ($\Delta f_{m,ads}$) and frequency shift as a consequence of viscosity and density changes ($\Delta f_{\eta,b(c)}$) in the electrolyte layer sensed by the EQCM.

$$\Delta f = \Delta f_{m,ads} + \Delta f_{\eta,b(c)} \tag{12}$$

The investigated corrosion inhibitors possess a good correlation between the adsorption data obtained from EQCM and the polarization curves [141]. While Akrout et al. used EQCM to discuss the adsorption of ascorbic acid salts (SA) and 1-hydroxyethylidene, 1-diphosphonic acid salts (SHEDP) by formation of chelates on steel which slow down the corrosion of the iron [142].

In addition, M. Scendo studied the inhibition properties of the purine (PU) and adenine (AD) on the corrosion of copper in 0.5 M Na_2SO_4 solutions and 1.0 M NaCl solutions by EQCM technique. The change in frequency is related to the mass variation due to the replaced water molecules per adsorbing inhibitor through the Sauerbrey equation discussed above. The surface coverage as obtained is very similar to the values derived from polarization curves [143,144].

Materials Research Forum LLC
https://doi.org/10.21741/9781644901052-4

Concluding remark

This chapter provides the effect of green corrosion inhibitors for preventing corrosion in micro to macro-devices. It also illuminates the general introduction about corrosion and its consequences on the economic and ecological development. Different forms and mechanism of corrosion have also been discussed with electrochemical monitoring techniques. Due to the wide span of corrosion prone materials and the challenging field conditions, corrosion cannot be fully prevented, but they can be detected, assessed, monitored, and managed. Also, exploration of green inhibitors in other corrosive medium should be analyzed such as carbon dioxide (CO_2) corrosion, Hydrogen Sulfide (H_2S) corrosion, and cooling water systems, which are at their infancy stage. However, it should be noted that the mentioned electrochemical techniques can complement each other for precise monitoring of corrosion owing to its stochastic nature of corrosion and large variation in field and operating conditions.

References

[1]   A.S.H. Makhlouf, Intelligent stannate-based coatings of self-healing functionality for magnesium alloys, in: A. Tiwari, J. Rawlins, L.H. Hihara (Eds.), Intelligent Coatings for Corrosion Control. Oxford, Butterworth Heinemann, 2015. pp. 537-555. https://doi.org/10.1016/B978-0-12-411467-8.00015-5

[2]   P. Zarras, J.D. Stenger-Smith, Smart inorganic and organic pretreatment coatings for the inhibition of corrosion on metals/alloys, in: A. Tiwari, J. Rawlins, L.H. Hihara (Eds.), Intelligent Coatings for Corrosion Control. Oxford, Butterworth Heinemann, 2015. pp. 59-91. https://doi.org/10.1016/B978-0-12-411467-8.00003-9

[3]   K. Y. Choi, S.S. Kim, Morphological analysis and classification of types of surface corrosion damage by digital image processing, Corros. Sci. 47 (2005) 1-15. https://doi.org/10.1016/j.corsci.2004.05.007

[4]   U.R. Evans, Stress corrosion: Its relation to other types of corrosion, Corrosion 7 (1951) 238-244. https://doi.org/10.5006/0010-9312-7.7.238

[5]   R.C. Ewing, Long-term storage of spent nuclear fuel, Nat. Mater. 14 (2015)252–257. https://doi.org/10.1038/nmat4226

[6]   W.J. Weber, RC. Ewing, C.R.A. Catlow, T.D. de la Rubia, L.W. Hobbs, C. Kinoshita, Hj. Matzke, A.T. Motta, M. Nastasi, E.K.H. Salje, E.R. Vance, S.J. Zinkle, Radiation effects in crystalline ceramics for the immobilization of high-level nuclear waste and plutonium, J. Mater. Res. 13 (1998) 1434-1484. https://doi.org/10.1557/JMR.1998.0205

[7]   H.H. Uhlig, The cost of corrosion to the United States, Corros. 6 (1950) 29-33. https://doi.org/10.5006/0010-9312-6.1.29

[8]   T.P. Hoar, Review lecture: Corrosion of metals: Its cost and control, Proc. R. Soc. Lond. A 348 (1976) 1-18. https://doi.org/10.1098/rspa.1976.0020

[9]   L.H. Bennett, J. Kruger, R.L. Parker, E. Passaglia, C. Reimann, A.W. Ruff, H. Yakowitz, E.B. Berman, Economic Effects of Metallic Corrosion in the United States: A Report to the Congress by the National Bureau of Standards; 1978. https://doi.org/10.6028/NBS.SP.511-1

[10]   M. V. Biezma, J.R. San Cristóbal, Methodology to study cost of corrosion, Corros. Eng. Sci. Techn., 40 (2005) 344–352. https://doi.org/10.1179/174327805X75821

[11]   B. Hou, X. Li, X. Ma, C. Du, D. Zhang, M. Zheng, M.W. Xu, D. Lu, F. Ma, The cost of corrosion in China, Mater. Degrad. 1 (2017) 1-10. https://doi.org/10.1038/s41529-017-0001-6

[12]   T.R. Witcher, From disaster to prevention: The silver bridge, Civ. Eng. 87 (2017) 44-47. https://doi.org/10.1061/ciegag.0001250

[13]   A.A. El-Meligi, Corrosion preventive strategies as a crucial need for decreasing environmental pollution and saving economics, Recent Pat. on Corros. Sci. 2 (2010) 22-33. https://doi.org/10.2174/1877610801002010022

[14]   P.P. Deshpande, N.G. Jadhav, V.J. Gelling, D. Sazou, Conducting polymers for corrosion protection: A review, J. Coat. Technol. Res., 11 (2014) 473–494. https://doi.org/10.1007/s11998-014-9586-7

[15]   S.A. Umoren, M.M. Solomon, Recent developments on the use of polymers as corrosion inhibitors: A review, Open Mater. Sci. J. 8 (2014) 39-54. https://doi.org/10.2174/1874088X01408010039

[16]   M. Taghavikish, N.K. Dutta, N.R. Choudhury, Emerging corrosion inhibitors for interfacial coating, Coatings 7 (2017) 217-245. https://doi.org/10.3390/coatings7120217

[17]   S.A. Umoren, U.F. Ekanem, Inhibition of mild steel corrosion in H_2SO_4 using exudate gum from pachylobus edulis and synergistic potassium halide additives, Chem. Eng. Comm. 197 (2010) 1339-1356. https://doi.org/10.1080/00986441003626086

[18]   C.G. Dariva, A.F. Galio, Corrosion inhibitors—principles, mechanisms and applications, in: M. Aliofkhazraei (Eds.), Developments in Corrosion Protection, Rijeka: InTech, 2014. pp. 365-380. https://doi.org/10.5772/57255

[19] M.N. Rahuma, B. Kannan, Corrosion in oil and gas industry: A perspective on corrosion inhibitors, J. Mater. Sci. Eng. 3 (2014) 110. https://doi.org/10.4172/2169-0022.1000e110

[20] N.O. Eddy, E.E. Ebenso, U.J. Ibok, Adsorption, synergistic inhibitive effect and quantum chemical studies of ampicillin (AMP) and halides for the corrosion of mild steel in H_2SO_4, J. Appl. Electrochem. 40 (2010) 445–456. https://doi.org/10.1007/s10800-009-0015-z

[21] A.S. Yaro, A.A. Khadom, R.K. Wael, Apricot juice as green corrosion inhibitor of mild steel in phosphoric acid, Alex. Eng. J. 52 (2013) 129-135. https://doi.org/10.1016/j.aej.2012.11.001

[22] R.M. Palou, O. Olivares-Xomelt, N.V. Likhanova, Environmentally friendly corrosion inhibitors, in: V.S. Sastri (1st Eds) Green Corrosion Inhibitors: Theory and Practice, Wiley, New Jersey, 2011, pp. 257-303. https://doi.org/10.1002/9781118015438.ch7

[23] I.B. Obot, N.O. Obi-Egbedi, S.A. Umoren, E.E. Ebenso, Synergistic and antagonistic effects of anions and ipomoea invulcrata as green corrosion inhibitor for aluminium dissolution in acidic medium, Int. J. Electrochem. Sci. 5 (2010) 994-1007.

[24] O. Gharbi, S. Thomas, C. Smith, N. Birbilis, Chromate replacement: What does the future hold?, Mater. Degrad. 2 (2018) 1-8. https://doi.org/10.1038/s41529-018-0034-5

[25] A.A.F. Sabirneeza, R. Geethanjali, S. Subhashini, Polymeric corrosion inhibitors for iron and its alloys: A review, Chem. Eng. Comm. 202 (2015) 232-244. https://doi.org/10.1080/00986445.2014.934448

[26] S.F. Hansen, L. Carlsen, J.A. Tickner, Chemicals regulation and precaution: Does REACH really incorporate the precautionary principle, Environ. Sci. Policy 10 (2007) 395-404. https://doi.org/10.1016/j.envsci.2007.01.001

[27] R.A.L. Sathiyanathan, S. Maruthamuthu, M. Selvanayagam, S. Mohanan, N. Palaniswamy, Corrosion inhibition of mild steel by ethanolic extracts of Ricinus communis leaves, Indian J. Chem. Technol. 12 (2005) 356-360.

[28] R.A.L. Sathiyanathan, M.M. Essa, S. Maruthamuthu, M. Selvanayagam, N. Palaniswamy, Inhibitory effect of Ricinus communis (castor-oil plant) leaf extract on corrosion of mild steel in low chloride medium, Indian J. Chem. Technol. 82 (2005) 357-359.

[29] N. Poongothai, P. Rajendran, M. Natesan, N. Palaniswamy, Wood bark oils as vapour phase corrosion inhibitors for metals in NaCl and SO_2 environments, Indian J. Chem. Technol. 12 (2005) 641-647.

[30] A.M. Abdel-Gaber, B.A. Abd-EL-Nabey, I.M. Sidahmed, A.M. El-Zayaday, M. Saadawy, Inhibitive action of some plant extracts on the corrosion of steel in acidic media, Corros. Sci. 48 (2006) 2765-2779. https://doi.org/10.1016/j.corsci.2005.09.017

[31] M. Kliskic, J. Radosevic, S. Gudic, V. Katalinic, Aqueous extract of Rosmarinus-o-cinalis L. as inhibitor of Al-Mg alloy corrosion in chloride solution, J. Appl. Electrochem. 30 (2006) 823-830. https://doi.org/10.1023/A:1004041530105

[32] A.N. Grassino, Z. Grabaric, A. Pezzani, G. Fasanaro, A. Lo Voi, Influence of essential onion oil on tin and chromium dissolution from tin-plate, Food Chem. Toxicol. 47 (2009) 1556-1561. https://doi.org/10.1016/j.fct.2009.04.003

[33] J. Halambek, K. Berkovic, J. Vorkapic-Furac, The influence of Lavandula angustifolia L. oil on corrosion of Al-3Mg alloy, Corros. Sci. 52 (2010) 3978-3983. https://doi.org/10.1016/j.corsci.2010.08.012

[34] P.C. Okafor, M.E. Ikpi, I.E. Uwah, E.E. Ebenso, U.J. Ekpe, S.A. Umoren, Inhibitory action of Phyllanthus amarus extracts on the corrosion of mild steel in acidic media, Corros. Sci. 50 (2008) 2310-2317. https://doi.org/10.1016/j.corsci.2008.05.009

[35] N.O. Eddy, Inhibitive and adsorption properties of ethanol extract of Colocasia esculenta leaves for the corrosion of mild steel in H_2SO_4, Int. J. Phys. Sci. 4 (2009)165-171. https://doi.org/10.1108/03699421011085849

[36] B. Hammouti, S. Kertit, M. Melhaoui, Electrochemical behaviour of bgugaine as a corrosion inhibitor of iron in 1 M HCl, Bull. Electrochem. 13 (1997) 97-98.

[37] A. Chetouani, B. Hammouti, M. Benkaddour, Corrosion inhibition of iron in hydrochloric acid solution by jojoba oil, Pigm. Resin Technol. 33 (2004) 26-31. https://doi.org/10.1108/03699420410512077

[38] E.L. Chaieb, A. Boyanzer, B. Hammouti, M. Benkaddour, M. Berrabah, Corrosion inhibition of steel in hydrpchloric acid solution by rosemary oil, Trans. SAEST 39 (2004) 58-60.

[39] E.L. Chaieb, A. Bouyanzer, B. Hammouti, M. Benkaddour, Inhibition of the corrosion of steel in 1 M HCl by eugenol derivatives, Appl. Surf. Sci. 246 (2005) 199-206. https://doi.org/10.1016/j.apsusc.2004.11.011

[40] O.S.I. Fayomi, P.A.L. Anawe, A. Daniyan, The impact of drugs as corrosion inhibitors on aluminum alloy in coastal-acidified medium, in: M. Aliofkhazraei (Eds.),

Corrosion Inhibitors, Principles and Recent Applications, Rijeka, InTech Open, 2018. pp. 79-94. https://doi.org/10.5772/intechopen.72942

[41]   A.H. Ali, Electrochemical study of candesartan drug as corrosion inhibitor for carbon steel in acid medium, J. Adv. Electrochem. 4 (2018) 152-157. https://doi.org/10.30799/jaec.050.18040101

[42]   P.B. Matad, P.B. Mokshanatha, N. Hebbar, V.T. Venkatesha, H.C. Tandon, Ketosulfone drug as a green corrosion inhibitor for mild steel in acidic medium, Ind. Eng. Chem. Res. 53 (2014) 8436-8444. https://doi.org/10.1021/ie500232g

[43]   P. Singh, D.S. Chauhan, K. Srivastava, V. Srivastava, M.A. Quraishi, Expired atorvastatin drug as corrosion inhibitor for mild steel in hydrochloric acid solution, Int. J. Ind. Chem. 8 (2017) 363-372. https://doi.org/10.1007/s40090-017-0120-5

[44]   B. El Ibrahimi, A. Jmiai, L. Bazzi, S. El Issami, Amino acids and their derivatives as corrosion inhibitors for metals and alloys, Arab. J. Chem. 13 (2020) 740-771. https://doi.org/10.1016/j.arabjc.2017.07.013

[45]   D.Q. Zhang, B. Xie, L.X. Gao, Q.R. Cai, H.G. Joo, K.Y. Lee, Intramolecular synergistic effect of glutamic acid, cysteine and glycine against copper corrosion in hydrochloric acid solution, Thin Solid Films 520 (2011) 356-361. https://doi.org/10.1016/j.tsf.2011.07.009

[46]   N.H. El-Sayed, Corrosion inhibition of carbon steel in chloride solutions by some amino acids, Eur. J. Chem. 7 (2016) 14-18. https://doi.org/10.5155/eurjchem.7.1.14-18.1331

[47]   M. Mobin, S. Zehra, M. Parveen, L-Cysteine as corrosion inhibitor for mild steel in 1 M HCl and synergistic effect of anionic, cationic and nonionic surfactants, J. Mol. Liq. 216 (2016) 598-607. https://doi.org/10.1016/j.molliq.2016.01.087

[48]   B.R.W. Hinton, N.E. Ryan, D.R. Arnott, P.N. Trathen, L. Wilson, B.E. Williams, The inhibition of aluminium alloy corrosion by rare earth metal cations, Corros. Austral. 10 (1985) 12.

[49]   B.R.W. Hinton, D.R. Arnott, N.E. Ryan, Cerium conversion coatings for the corrosion protection of aluminum, Mater. Forum 9 (1986) 162.

[50]   B.R.W. Hinton, D.R. Arnott, N.E. Ryan, The inhibition of aluminum alloy corrosion by cerous cations, Metals Forum 7 (1984) 11.

[51]   M. Bethencourt, F.J. Botana, J.J. Calvino, M. Marcos, M.A. Rodríguez-Chacón, Lanthanide compounds as environmentally-friendly corrosion inhibitors of aluminium alloys: A review, Corros. Sci. 40 (1998) 1803-1819. https://doi.org/10.1016/S0010-938X(98)00077-8

[52] A.K. Bhattamishra, M.K. Banerjee, Corrosion behavior of Al-Zn-Mg alloys in NaCl solution in presence of cerium salts, Zeitschrift fur metallkunde, 84 (1993) 734-736.

[53] B.R.W. Hinton, Corrosion inhibition with rare earth metal salts, J. Alloys Compd. 180 (1992) 15-25. https://doi.org/10.1016/0925-8388(92)90359-H

[54] T. Zhang, D.Y.Li, The effect of YCl_3 and $LaCl_3$ additives on wear of 1045 and 304 steels in a dilute chloride solution, Mater. Sci. Eng.: A 345 (2003) 179-189. https://doi.org/10.1016/S0921-5093(02)00469-0

[55] O. Lopez-Garrity, G.S. Frankel, Corrosion Inhibition of Aluminum Alloy 2024-T_3 by Praseodymium Chloride, Corros. 70 (2014) 928-941. https://doi.org/10.5006/1244

[56] M.J. Bennett, B.A. Bellamy, G. Dearnaley, M.R. Houlton, The influence of Eu, La, Sc, Yb, ion-implantation upon the oxidation behaviour of a 20Cr25NiNb stabilized stainless steel in carbon dioxide at 825° C, Proc. Int. Cong. Metal Corros. 2 (1984) 416-423.

[57] M. Bethencourt, F.J. Botana, J.J. Calvino, M. Marcos, M. A. Rodrigues-Chacon, Lanthanide compounds as environmentally-friendly corrosion inhibitors of aluminium alloys: a review, Corros. Sci. 40 (1998) 1803-1819. https://doi.org/10.1016/S0010-938X(98)00077-8

[58] L. Guo, S. Zhu, S. Zhang, Experimental and theoretical studies of benzalkonium chloride as an inhibitor for carbon steel corrosion in sulfuric acid, J. Ind. Eng. Chem. 24 (2015) 174-180. https://doi.org/10.1016/j.jiec.2014.09.026

[59] M.A. Hegazy, A.Y. El-Etre, M. El-Shafaie, K.M. Berry, Novel cationic surfactants for corrosion inhibition of carbon steel pipelines in oil and gas wells applications, J. Mol. Liq. 214 (2016) 347–356. https://doi.org/10.1016/j.molliq.2015.11.047

[60] Y. Zhu, M.L. Free, J.H. Cho, Integrated evaluation of mixed surfactant distribution in water-oil-steel pipe environments and associated corrosion inhibition efficiency, Corros. Sci. 110 (2016) 213–227. https://doi.org/10.1016/j.corsci.2016.04.043

[61] F.E.T. Heakal, A.Y. Elkholy, Gemini surfactants as corrosion inhibitors for carbon steel, J. Mol. Liq. 230 (2017) 395–407. https://doi.org/10.1016/j.molliq.2017.01.047

[62] S.M. Shaban, R.M. El-Sherif, M.A. Fahim, Studying the surface behavior of some prepared free hydroxyl cationic amphipathic compounds in aqueous solution and their biological activity, J. Mol. Liq. 252 (2018) 40–51. https://doi.org/10.1016/j.molliq.2017.12.105

[63] S.M. Shaban, A.A. Abd-Elaal, S.M. Tawfik, Gravimetric and electrochemical evaluation of three nonionic dithiol surfactants as corrosion inhibitors for mild steel in

1 M HCl solution, J. Mol. Liq. 216 (2016) 392–400.
https://doi.org/10.1016/j.molliq.2016.01.048

[64] Z. Zhang, N. Tian, X. Li, L. Zhang, L. Wu, Y. Huang, Synergistic inhibition behavior between indigo carmine and cetyl trimethyl ammonium bromide on carbon steel corroded in a 0.5 M HCl solution, Appl. Surf. Sci. 357 (2015) 845–855.
https://doi.org/10.1016/j.apsusc.2015.09.092

[65] M. Mobin, R. Aslam, S. Zehra, M. Ahmad, Bio-/Environment-Friendly Cationic Gemini Surfactant as Novel Corrosion Inhibitor for Mild Steel in 1 M HCl Solution, J. Surfactants Deterg. 20 (2017) 57–74. https://doi.org/10.1007/s11743-016-1904-x

[66] X. Chen, X. Li, A. Hu, F. Wang, Advances in chiral ionic liquids derived from natural amino acids, Tetrahedron: Asymmetry 19 (2008) 1-14.
https://doi.org/10.1016/j.tetasy.2007.11.009

[67] M. Hasib-ur-Rahman, M. Siaj, F. Larachi, Ionic liquids for CO_2 capture—development and progress, Chem. Eng. Process. 49 (2010) 313-322.
https://doi.org/10.1016/j.cep.2010.03.008

[68] M.E. Mashuga, L.O. Olasunkanmi, A.S. Adekunle, S. Yesudass, M.M. Kabanda, E.E. Ebenso, Adsorption, thermodynamic and quantum chemical studies of 1-hexyl-3-methylimidazolium based ionic liquids as corrosion inhibitors for mild steel in HCl, Mater. 8 (2015) 3607-3632. https://doi.org/10.3390/ma8063607

[69] M.C. Bubalo, K. Radošević, I.R. Redovniković, J. Halambek, V.G. Srček, A brief overview of the potential environmental hazards of ionic liquids, Ecotoxicol. Environ. Saf. 99 (2014) 1-12. https://doi.org/10.1016/j.ecoenv.2013.10.019

[70] M.P. Singh, R.K. Singh, S. Chandra, Ionic liquids confined in porous matrices: physicochemical properties and applications, Prog. Mater. Sci. 64 (2014) 73-120.
https://doi.org/10.1016/j.pmatsci.2014.03.001

[71] S.M. Tawfik, Ionic liquids based gemini cationic surfactants as corrosion inhibitors for carbon steel in hydrochloric acid solution, J. Mol. Liq. 216 (2016) 624-635. https://doi.org/10.1016/j.molliq.2016.01.066

[72] R.K. Blundell, P. Licence, Quaternary ammonium and phosphonium based ionic liquids: a comparison of common anions, Phys. Chem. Chem. Phys. 16 (2014) 15278-15288. https://doi.org/10.1039/C4CP01901F

[73] S.M. Tawfik, Ionic liquids based Gemini cationic surfactants as corrosion inhibitors for carbon steel in hydrochloric acid solution, J. Mol. Liq. 216 (2016) 624-635. https://doi.org/10.1016/j.molliq.2016.01.066

[74]   H.H. Elsentriecy, H. Luo, H.M. Meyer, L.L. Grado, J. Qu, Effects of pretreatment and process temperature of a conversion coating produced by an aprotic ammonium-phosphate ionic liquid on magnesium corrosion protection, Electrochim. Acta 123 (2014) 58-65. https://doi.org/10.1016/j.electacta.2013.12.167

[75]   R.G. Kelly, Electrochemical Thermodynamics and Kinetics of Relevance to Corrosion, in: R.G. Kelly, J.R. Scully, D.W. Shoesmith, R.G. Buchheit, Electrochemical Techniques in Corrosion Science and Engineering, Marcel Dekker Inc., 2002, pp. 9-54. https://doi.org/10.1201/9780203909133

[76]   C.N. Chang, H.B. Cheng, A.C. Chao, Applying the Nernst Equation to simulate redox potential variations for biological nitrification and denitrification processes, Environ. Sci. Technol. 38 (2004) 1807-1812. https://doi.org/10.1021/es021088e

[77]   K.B. Oldham, F. Mansfeld, On the so-called linear polarization method for measurement of corrosion rates, Corros. 27 (1971) 434-435. https://doi.org/10.5006/0010-9312-27.10.434

[78]   A. Goyal, H.S. Pouya, E. Ganjian, A.O. Olubanwo, M. Khorami, Predicting the corrosion rate of steel in cathodically protected concrete using potential shift, Constr. Build Mater. 194 (2019) 344-349. https://doi.org/10.1016/j.conbuildmat.2018.10.153

[79]   S.W. Dean, Corrosion monitoring for industrial processes, in: DS Cramer, BS Covino (Eds.), Corrosion: Fundamentals, Testing and Protection, Metals Park, OH, ASM International, 2003, pp. 533–541. https://doi.org/10.31399/asm.hb.v13a.a0003659

[80]   S. Papavinasam, R.W. Revie, M. Attard, A. Demoz, K. Michaelian, Comparison of techniques for monitoring corrosion inhibitors in oil and gas pipelines, Corrosion 59 (2003) 1096-1111. https://doi.org/10.5006/1.3277529

[81]   F. Suedile, F. Robert, C. Roos, M. Lebrini, Corrosion inhibition of zinc by Mansoa alliacea plant extract in sodium chloride media: Extraction, Characterization and Electrochemical Studies, Electrochim. Acta 133 (2014) 631-638. https://doi.org/10.1016/j.electacta.2013.12.070

[82]   (a) J. Tafel, On the polarization during cathodic hydrogen evolution, Z. Phys. Chem 50 (1905) 641. (b) X.L. Zhang, Z.H. Jiang, Z.P. Yao, Y. Song, Z.D. Wu, Effects of scan rate on the potentiodynamic polarization curve obtained to determine the Tafel slopes and corrosion current density, Corros. Sci. 51 (2009) 581-587. https://doi.org/10.1016/j.corsci.2008.12.005

[83]   R. Kanojia, G. Singh, An interesting and efficient organic corrosion inhibitor for mild steel in acidic medium, Surf. Eng. 21 (2005) 180-186. https://doi.org/10.1179/174329405X49985

[84]   O. E. Herrera, D.P. Wilkinson, W. Merida, Anode and cathode overpotentials and temperature profiles in a PEMFC, J. Power Sources 198 (2012) 132-142. https://doi.org/10.1016/j.jpowsour.2011.09.042

[85]   E. McCafferty, Validation of corrosion rates measured by the Tafel extrapolation method, Corros. Sci. 47 (2005) 3202-3215. https://doi.org/10.1016/j.corsci.2005.05.046

[87]   H. Liu, Y.X. Leng, G. Wan, N. Huang, Corrosion susceptibility investigation of Ti–O film modified cobalt-chromium alloy (L-605) vascular stents by cyclic potentiodynamic polarization measurement, Surf. Coat. Tech. 206 (2011) 893-896. https://doi.org/10.1016/j.surfcoat.2011.04.048

[88]   S. Esmailzadeh, M. Aliofkhazraei, H. Sarlak, Interpretation of cyclic Potentiodynamic polarization test results for study of corrosion behavior of metals: A review, Prot. Met. Phys. Chem. 54 (2018) 976–989. https://doi.org/10.1134/S207020511805026X

[89]   M.A. Deyab, Corrosion inhibition of aluminum in biodiesel by ethanol extracts of Rosemary leaves, J. Taiwan Inst. Chem. E. 58 (2016) 536-541. https://doi.org/10.1016/j.jtice.2015.06.021

[90]   M. Kliskic, J. Radoservic, S. Gudic, V. Katalinic, Aqueous extract of Rosmarinus officinalis L. as inhibitor of AlMg alloy corrosion in chloride solution, J. Appl. Electrochem. 30 (2000) 823–830. https://doi.org/10.1023/A:1004041530105

[91]   A. Singh, I. Ahamad, M.A. Quraishi, Piper longum extract as green corrosion inhibitor for aluminium in NaOH solution, Arab. J. Chem. 9 (2016) S1584-S1589. https://doi.org/10.1016/j.arabjc.2012.04.029

[92]   Z. Tao, S. Zhang, W. Li, B. Hou, Adsorption and inhibitory mechanism of 1H-1,2,4-triazol-l-yl-methyl-2-(4-chlorophenoxy) acetate on corrosion of mild steel in acidic solution, Ind. Eng. Chem. Res. 50 (2011) 6082-6088. https://doi.org/10.1021/ie101793b

[93]   A. Michel, B.J. Pease, M.R. Geiker, H. Stang, J.F. Olesen, Monitoring reinforcement corrosion and corrosion-induced cracking using non-destructive x-ray attenuation measurements, Cement Concrete Res. 41 (2011) 1085-1094. https://doi.org/10.1016/j.cemconres.2011.06.006

[94]   R. Barker, B. Pickles, N. Kapur, T. Hughes, E. Barmatov, A. Neville, Flow cell apparatus for quantitative evaluation of carbon steel corrosion during transitions in fluid composition: Application to transition from inhibited hydrochloric acid to sodium chloride brine, Corros. Sci., 2018, 138, 116-129. https://doi.org/10.1016/j.corsci.2018.04.012

[95] W. Ding, M. Li, M. Wang, R. Chen, Y. Wang, L. Chen, Experimental study on corrosion of anchored rock mass for half-through intermittent joints, Adv. Civ. Eng. 12 (2019) 6018678. https://doi.org/10.1155/2019/6018678

[96] S. Gao, B. Brown, D. Young, M. Singer, Formation of iron oxide and iron sulfide at high temperature and their effects on corrosion, Corros. Sci. 135 (2018) 167-176. https://doi.org/10.1016/j.corsci.2018.02.045

[97] D.V. Ribeiro, J.C.C. Abrantes, Application of electrochemical impedance spectroscopy (EIS) to monitor the corrosion of reinforced concrete: A new approach, Constr. Build. Mater. 111 (2016) 98-104. https://doi.org/10.1016/j.conbuildmat.2016.02.047

[98] P. Ahuja, S.K. Ujjain, R. Kanojia, MnO_x/C nanocomposite: An insight on high-performance supercapacitor and non-enzymatic hydrogen peroxide detection, Appl. Surf. Sci. 404 (2017) 197-205. https://doi.org/10.1016/j.apsusc.2017.01.300

[99] P. Ahuja, S.K. Ujjain, R. Kanojia, Electrochemical behaviour of manganese & ruthenium mixed oxide@ reduced graphene oxide nanoribbon composite in symmetric and asymmetric supercapacitor, Appl. Surf. Sci. 427 (2018) 102-111. https://doi.org/10.1016/j.apsusc.2017.08.028

[100] S.K. Ujjain, P. Ahuja, R. Bhatia, P. Attri, Printable multi-walled carbon nanotubes thin film for high performance all solid state flexible supercapacitors, Mater. Res. Bull. 83 (2016) 167-171. https://doi.org/10.1016/j.materresbull.2016.06.006

[101] S. M. Hoseinieh, A.M. Homborg, T. Shahrabi, J.M.C. Mol, B. Ramezanzadeh, A Novel Approach for the evaluation of under deposit corrosion in marine environments using combined analysis by electrochemical impedance spectroscopy and electrochemical noise, Electrochim. Acta 217 (2016) 226-241. https://doi.org/10.1016/j.electacta.2016.08.146

[102] M.F. Montemor, A.M.P. Simoes, M.G.S. Ferreira, Chloride-induced corrosion on reinforcing steel: from the fundamentals to monitoring techniques, Cement Concr. Compos. 25 (2003) 491-502. https://doi.org/10.1016/S0958-9465(02)00089-6

[103] P. Langford, J. Broomfield, Monitoring the corrosion of reinforcing steel, Constr. Repair 1 (1987) 32-36.

[104] A. Aguilar, A. Sagüés, R. Powers, Corrosion measurements of reinforcing steel in partially submerged concrete slabs, in: N. Berke, V. Chaker, D. Whiting (Eds.) Corrosion Rates of Steel in Concrete, West Conshohocken, PA: ASTM International, 1990, pp. 66-85. https://doi.org/10.1520/STP25016S

[105] M. Lebrini, F. Suedile, C. Roos, Corrosion inhibitory action of ethanol extract from Bagassa guianensis on the corrosion of zinc in ASTM medium, J. Mater. Environ. Sci. 9 (2018) 414-423.

[106] R. Fdil, M. Tourabi, S. Derhali, A. Mouzdahir, K. Sraidi, C. Jama, A. Zarrouk, F. Bentiss, Evaluation of alkaloids extract of Retama monosperma (L.) Boiss. stems as a green corrosion inhibitor for carbon steel in pickling acidic medium by means of gravimetric, AC impedance and surface studies, J. Mater. Environ. Sci. 9 (1) (2018) 358-369. https://doi.org/10.26872/jmes.2018.9.1.39

[107] H. Grengi, H.I. Sahin, Schinopsis Lorentzii extract as a green corrosion inhibitor for low carbon steel in 1 M HCl solution, Ind. Eng. Chem. Res. 51 (2012) 780–787. https://doi.org/10.1021/ie201776q

[108] A.M. Abdel-Gaber, B.A. Abd-El-Nabey, I.M. Sidahmed, A.M. El-Zayaday, M. Saadawy, Inhibitive action of some plant extracts on the corrosion of steel in acidic media, Corros. Sci. 48 (2006) 2765–2779. https://doi.org/10.1016/j.corsci.2005.09.017

[109] L.R. Chauhan, G. Gunasekaran, Corrosion inhibition of mild steel by plant extract in dilute HCl medium, Corros. Sci. 49 (2007) 1143-1161. https://doi.org/10.1016/j.corsci.2006.08.012

[110] M. Lebrini, F. Robert, A. Lecante, C. Roos, Corrosion inhibition of C38 steel in 1 M hydrochloric acid medium by alkaloids extract from Oxandra asbeckii plant, Corros. Sci. 53 (2011) 687-695. https://doi.org/10.1016/j.corsci.2010.10.006

[111] D.G. John, P.C. Searson, J.L. Dawson, Use of AC Impedance technique in studies on steel in concrete in immersed conditions, Br. Corros. J. 16 (1981) 102-106. https://doi.org/10.1179/000705981798275002

[112] L. Dhouibi-Hachani, E. Triki, J. Grandet, A. Raharinaivo, Comparing the steel-concrete interface state and its electrochemical impedance, Chem. Concr. Res. 26 (1996) 253-266. https://doi.org/10.1016/0008-8846(95)00214-6

[113] A.A. Sagues, S.C. Kranc, E.L. Moreno, The time domain response of corroding systemwith constant phase angle interfacial component: application to steel in concrete, Corros. Sci. 37 (1995) 1097-1113. https://doi.org/10.1016/0010-938X(95)00017-E

[114] V. Feliu, J.A. González, C. Andrade, S. Feliu, Equivalent circuit for modelling the steel-concrete interface I: Experimental evidence and theoretical predictions, Corros. Sci. 40 (1998) 975-993. https://doi.org/10.1016/S0010-938X(98)00036-5

[115] S.K. Ujjain, P. Ahuja, R.K. Sharma, Graphene nanoribbon wrapped cobalt manganite nanocubes for high performance all-solid-state flexible supercapacitors, J. Mater. Chem. A 3 (2015) 9925-9931. https://doi.org/10.1039/C5TA00653H

[116] S.K. Ujjain, R. Bhatia, P. Ahuja, P. Attri, Highly conductive aromatic functionalized multi-walled carbon nanotube for inkjet printable high performance supercapacitor electrodes, Plos One 10 (2015) e0131475. https://doi.org/10.1371/journal.pone.0131475

[117] B.J. Christensen, T. Coverdale, R.A. Olson, S.J. Ford, E.J. Garboczi, H.M. Jennings, T.O. Mason, Impedance Spectroscopy of hydrating cement-based materials: measurement, interpretation and application, J. Am. Ceram. Soc. 77 (1994) 2789-2804. https://doi.org/10.1111/j.1151-2916.1994.tb04507.x

[118] B.J. Christensen, T.O. Mason, H.M. Jennings, Influence of silica fume on early hydration of Portland cements using impedance spectroscopy, J. Am. Ceram. Soc. 75 (1992) 939-945. https://doi.org/10.1111/j.1151-2916.1992.tb04163.x

[119] A. Legat, V. Doleček, Corrosion monitoring system based on measurement and analysis of electrochemical noise, Corros. 51 (1995) 295-300. https://doi.org/10.5006/1.3293594

[120] J. Dawson, Electrochemical noise measurement: The definitive in-situ technique for corrosion applications, in: J. Kearns, J. Scully, P. Roberge, D. Reichert, J. Dawson (Eds.) Electrochemical Noise Measurement for Corrosion Applications, West Conshohocken, PA, ASTM International, 1996, pp. 3-35.

[121] R. A. Cottis, Interpretation of electrochemical noise data, Corros. 57 (2001) 265-285. https://doi.org/10.5006/1.3290350

[122] (a) U. Bertocci, Separation between deterministic response and random fluctuations by means of the cross-power spectrum in the study of electrochemical noise, J. Electrochem. Soc. 128 (1981) 520-523. (b) A.M.P. Simoes, M.G. S. Ferreira, Crevice corrosion studies on stainless steel using electrochemical noise measurements, Brit. Corros. J. 22 (1987) 21-25. https://doi.org/10.1179/000705987798271802

[123] P.C. Searson, J.L. Dawson, Analysis of electrochemical noise generated by corroding electrodes under open-circuit conditions, J. Electrochem. Soc. 135 (1988) 1908-1915. https://doi.org/10.1149/1.2096177

[124] K. Hladky, J.L. Dawson, The measurement of localized corrosion using electrochemical noise, Corros. Sci. 21 (1981) 317-322. https://doi.org/10.1016/0010-938X(81)90006-8

[125] A. Ehsani, M.G. Mahjani, M. Hosseini, R. Safari, R. Moshrefi, H.M. Shiri, Evaluation of Thymus vulgaris plant extract as an eco-friendly corrosion inhibitor for stainless steel 304 in acidic solution by means of electrochemical impedance spectroscopy, electrochemical noise analysis and density functional theory, J. Colloid Interface Sci. 490 (2017) 444–451. https://doi.org/10.1016/j.jcis.2016.11.048

[126] M. Mehdipour, B. Ramezanzadeh, S.Y. Arman, Electrochemical noise investigation of Aloe plant extract as green inhibitor on the corrosion of stainless steel in 1 M H_2SO_4, J. Ind. Eng. Chem. 21 (2015) 318–327. https://doi.org/10.1016/j.jiec.2014.02.041

[127] I.A. Hermoso-Diaz, M.A. Velázquez-González, M.A. Lucio-Garcia, J.G. Gonzalez-Rodriguez, A study of Salvia hispanica as green corrosion inhibitor for carbon steel in sulfuric acid, Chem. Sci. Rev. Lett. 3 (2014) 685-697.

[128] M. Shahidi, E. Sasaei, M. Ganjehkaviri, M.R. Gholamhosseinzadeh, Investigation of the effect of vanillin as a green corrosion inhibitor for stainless steel using electrochemical techniques, J. Phys. Theor. Chem. IAU Iran 9 (2012) 149-161.

[129] A.M. Nagiub, Electrochemical noise analysis of different herbal compounds for copper exposed to chloride media, Int. J. Electrochem. Sci., 11 (2016) 7861 – 7874. https://doi.org/10.20964/2016.09.19

[130] B. Ramezanzadeh, S.Y. Arman, M. Mehdipour, B.P. Markhali, Analysis of electrochemical noise (ECN) data in time and frequency domain for comparison corrosion inhibition of some azole compounds on Cu in 1.0 M H_2SO_4 solution, Appl. Surf. Sci. 289 (2014) 129-140. https://doi.org/10.1016/j.apsusc.2013.10.119

[131] S.P. Sharma, Reaction of Copper and Copper Oxide with H_2S, J. Electrochem. Soc., 127 (1980) 21-26. https://doi.org/10.1149/1.2129622

[132] M. Itagaki, H. Nakazawa, K. Watanabe, K. Noda, Study of dissolution mechanisms of nickel in sulfuric acid solution by electrochemical quartz crystal microbalance, Corros. Sci. 39 (1997) 901-911. https://doi.org/10.1016/S0010-938X(97)81157-2

[133] M. Fonsati, F. Zucchi, G. Trabanelli, Study of corrosion inhibition of copper in 0.1 M NaCl using the EQCM technique, Electrochim. Acta 44 (1998) 311-322. https://doi.org/10.1016/S0013-4686(98)00170-4

[134] P. Curie, J. Curie, Electric field induced contractions and expansions in hemihedral crystals with inclined faces, CR Acad. Sci. Paris 91 (1880) 294-297.

[135] V.S. Muralidharan, Critical review on electrochemical quartz crystal micro balance-principles and applications to corrosion research, Bull. Electrochem. 17 (2001) 183-192.

[136] A. Ispas, E. Wolff, A. Bund, An electrochemical quartz crystal microbalance study on electrodeposition of aluminum and aluminum-manganese alloys, J. Electrochem. Soc, 164 (2017) 5263-5270. https://doi.org/10.1149/2.0381708jes

[137] E.M. Moustafa, S. Zein El Abedin, A. Shkurankov, E. Zschippang, A.Y. Saad, A. Bund, F. Endres, Electrodeposition of Al in 1-Butyl-1-methylpyrrolidinium Bis(trifluoromethylsulfonyl)amide and 1-Ethyl-3-methylimidazolium Bis(trifluoromethylsulfonyl)amide Ionic Liquids: In Situ STM and EQCM Studies, J. Phys. Chem. B 111 (2007) 4693-4704. https://doi.org/10.1021/jp0670687

[138] V.T. Gruia, A. Ispas, M. Wilke, I. Efimov, A. Bund, Application of acoustic impedance method to monitoring of sensors: Metal deposition on viscoelastic polymer substrate, Electrochim. Acta 118 (2014) 88-91. https://doi.org/10.1016/j.electacta.2013.12.010

[139] S. Ivanov, C. Vlaic, A. Bund, Igor Efimov, In situ analysis of surface morphology and viscoelastic effects during deposition of thin silicon layers from 1-butyl-1-methylpyrrolidinium bis(trifluoromethylsulfonyl)imide, Electrochim. Acta 219 (2016) 251-257. https://doi.org/10.1016/j.electacta.2016.09.156

[140] C.O.A. Olsson, D. Landolt, Electrochemical quartz crystal microbalance, in: P. Marcus, F. Mansfeld, (Eds.) Analytical Methods in Corrosion Science and Engineering. CRC Press, Taylor and Francis Group, Boca Raton, Florida, 2006, pp. 733-751. https://doi.org/10.1201/9781420028331.ch19

[141] P. Kern, D. Landolt, Adsorption of organic corrosion inhibitors on iron in the active and passive state. A replacement reaction between inhibitor and water studied with the rotating quartz crystal microbalance, Electrochim. Acta 47 (2001) 589–598. https://doi.org/10.1016/S0013-4686(01)00781-2

[142] H. Akrout, L. Bousselmi, E. Triki, S. Maximovitch, F. Dalard, Adsorption mechanism of non-toxic organic inhibitors on steel in solutions at pH 8 determined by electrochemical quartz crystal microbalance measurements, Mater. Corros. 56 (2005) 185-191. https://doi.org/10.1002/maco.200403828

[143] M. Scendo, Inhibitive action of the purine and adenine for copper corrosion in sulphate solutions, Corros. Sci. 49 (2007) 2985–3000. https://doi.org/10.1016/j.corsci.2007.01.002

[144] M. Scendo, The effect of purine on the corrosion of copper in chloride solutions, Corros. Sci. 49 (2007) 373–390. https://doi.org/10.1016/j.corsci.2006.06.022

Theory and Applications of Green Corrosion Inhibitors   Materials Research Forum LLC
Materials Research Foundations **86** (2021) 127-160   https://doi.org/10.21741/9781644901052-5

Chapter 5

Green Corrosion Inhibitors for Technological Applications

P. Jain[1], S. Raghav[2], D. Kumar[3]*

[1]Department of Chemistry, SRM-IST, Delhi-NCR Campus, Modinagar-210204, India

[2]Department of Chemistry, Banasthali Vidyapith, Banasthali, Tonk 304022, India

[3]School of Chemical Sciences, Central University of Gujarat, Gandhinagar, India

* dinesh.kumar@cug.ac.in

Abstract

Corrosion is an inevitable fact of day-to-day life, and however, because of its technological, economic, and aesthetic significance, it always receives much attention. Most of the corrosion inhibitors are environmentally harmful and toxic synthetic chemicals. In view of the toxicity of the inhibitors, the search for an eco-friendly and non-toxic corrosion inhibitor is of great interest. Green corrosion inhibitors are of concern because of increased awareness and improvements in regulations related to the environment because of their toxicity, restrict regular corrosion inhibitors. The extracts of natural products contain compounds having oxygen, carbon, nitrogen and sulfur. Such elements facilitate compounds to absorb on the surface of metal, forming a protective film to prevent corrosion. The main purpose of this chapter is to provide a comprehensive study of technological applications of green corrosion inhibitors in different industries, such as reinforced concrete, coating, aircraft, oil and gas, acid pickling, and water industry.

Keywords

Corrosion, Green Inhibitors, Eco-Friendly, Non-Toxic, Natural Products, Applications

Contents

1.   Introduction

Corrosion is the chemical attack or reaction of a metal with its atmosphere and is impossible to remove entirely [1-5]. Corrosion affects the quality of the environment, the efficiency of industries, and the assets of infrastructure [6]. The pollutants speed up the corrosion and form undesirable products such as rust, oxides, etc. The financial burden of corrosion causes countries to monitor and reduce the process of corrosion, every year [7-10]. Due to the low cost, durability, and mechanical properties of mild steel, it is used in various industrial applications but undergoes corrosion in various environments [11-15]. The use of corrosion inhibitors [16-20] is very common among several methods of corrosion control, i.e., cathodic protection, anodic protection, protective coatings, and material selection [21]. Any substance that is added (small quantity) to a corrosive environment to minimize or prevent corrosion is referred to as a corrosion inhibitor [22-28]. The most well-known corrosion inhibitors are heteroatoms containing organic compounds and multiple bonds, which is because of high electron density and basicity [9-11]. Basically, a protective film is formed on the metal surfaces, which acts as a barrier and inhibits corrosion [29]. Corrosion inhibitor safety and environmental concerns have always been a global problem in industries. Several synthetic compounds have shown good corrosion inhibition properties but are costly and highly poisonous to humans and the environment [30]. Various plant extracts (natural products) have been exposed as green corrosion inhibitors, as these are environmentally friendly, less costly, easier to access, reusable, and potential [31-37]. This plant extract contains high amounts of natural chemical compounds (π-electron centers) that can be processed through simple, low-cost procedures and easily absorbed on the metal surface [38-45].

2.   Green corrosion inhibitors

The concept of green chemistry the science field, limits the use of traditional inhibitors because of increased ecological concern and environment-friendly regulations [46-50]. Green chemistry is a branch of science with the goal of designing eco-friendly chemicals to reduce the release of harmful chemicals/materials into the environment [51-54]. This goal inspired the scientists/engineers to synthesis the organic corrosion inhibitors involving multi-component reactions in a single step. These multi-component reactions are a very significant tool for the green synthesis of organic compounds, especially in combination with ultrasound and microwave irradiations [55-58].

In addition, various plant extracts [59,60], chemical medicinal products [61,62], and ionic liquids [63,64] have also been employed to produce green corrosion inhibitors. Several types of research have been already reported, such as green corrosion inhibitors independent of chromate [65,66], organic inhibitors [67,68], inorganic inhibitors [69,70], and many more. A lot of work has been performed over the years to study these natural resources [71,72]. For hydrochloric acid solutions, Delonixregia used rosemary leaves as a green blocker to protect aluminum (Al) [73]. The other research [74] demonstrated corrosion inhibition on mild steel with African bush pepper extract (*Piper guinensis*). The extract of black pepper (*piperaceae*) and nicotine showed excellent efficiency as corrosion inhibitors (CIs). Black pepper extract gave 98% efficiency of inhibition for mild steel in HCl at 120 ppm at 35 °C [75]. The electrochemical analysis discovered that it is a mixed-type inhibitor that regulates the corrosion process [76].

3.   Technological applications of green corrosion inhibitors

3.1   Oil and gas sector

Generally, pipelines are buried underground and used to transport liquids and gases worldwide to final users over long distances from their sources [77]. Corrosion has resulted in the malfunction of pipelines expensive to repair [78,79]. In oil and gas production, corrosion is mainly because of fluids in pipelines, carbon dioxide, and hydrogen sulfide gases in contact with water [80,81]. Various diverse processes are used to prevent and manage the corrosion of gas and oil pipelines including advancement of the construction material, cathodic protection, application of the protective coating, and use of inhibitors for the modification of the corrosive environment. Steel (Iron) and its alloys are extensively employed in an oil refinery equipment, tanks, pipelines, etc.

construction because of its high strength. The aggressive reaction in acidic solutions is the main problem of using steel alloys [82]. Acid solutions are widely used in many industrial processes to take out unwanted rust and scale. Hydrochloric acid is one acid that is commercially available. The corrosion rate of steel in liquid media enhances with increased rate of flow because of the mechanical degradation of the passive films and dissolution of the growth of corrosion products, at various flow rates. The typical method to manage the dissolution of steel and alloys, besides the consumption of acid, is the addition of inhibitors [83]. By using the inhibitors, the metal consumption reaction rate is modified in acids as well as affects the kinetics of the electrochemical reactions. The use of some CIs, such as chromate, was forbidden owing to toxicity and environmental hazards [84]. Thus, eco-friendly, non-hazardous, natural plant extracts are used as corrosion inhibitors and known as green corrosion inhibitors. Plant extracts contain a broad range of heteroatoms containing organic compounds (S, O, N, and P). Extracts from mango, orange, banana, aloe vera, tobacco leaves etc. are some organic corrosion inhibitors used in the past [85-87].

The property of corrosion inhibition was attributed to "piperine" alkaloid. Nicotine is an organic compound of the alkaloids, colorless arginine derivative. Nicotine provides safety to the petroleum pipeline systems. It is found readily in tobacco plants (*Nicotiana tobacum*), and is the main active chemical element. In addition, the extract of *the Nicotiana tabacum* showed that the best concentration was 1200 mg/L with an efficiency of 89% [88]. Whereas in other research, it was found that inhibitory corrosion properties extract of *nicotiana tabacum* in acidic steel reached 94.13% at the concentration of 10 g/L for 6 hours at 303 K under static conditions [89].

Eleutherine americanna Merr. (Bawang Dayak extract) was listed as an environment-friendly green organic inhibitor of mixed type with flavonoid compounds that took part in the corrosion inhibitor cycle on API 5L X42 steel. The maximum efficiency of inhibition, i.e., 84.5% occurred when the concentration of 1000 ppm inhibitor is added to 1M HCl. At 1000 ppm, the corrosion current density of steel decreased to 81635 µA, and solution resistance increased to 748 Ω. Bawang Dayak's extract inhibited the isothermic Langmuir mode, which will generally control the process of charge transfer on the steel interface of the API 5L X42 [90]. Another extract of *Brassica oleracea* demonstrated outstanding mixed-type corrosion prevention properties in acidic conditions, i.e., 0.5M H_2SO_4 of pipeline steel corrosion. *Brassica oleracea* extract's anticorrosion actions are because of its multi constituent. The subsequent association of *Brassica oleracea's* multi constituent accounts for its high-performance inhibition. *Brassica oleracea* inhibition was caused by the development of an adsorbed corrosion retarding system of Iron and molecular components in *Brassica oleracea* and an accelerated complex of unstable soluble corrosion [91]. Castor oil, Neem oil, and punga oil also showed a green corrosion inhibitor (mixed type) in the petroleum pipeline (API 5LX) used for diesel transportation. These oils were evaluated along with water (2%) containing chloride (120 ppm). Weight loss and polarization analysis were done to determine the effectiveness of corrosion inhibition. The study of weight loss revealed that castor oil gave higher efficiency (89%) compared to other punga oil and neem oil [92].

3.2   Reinforced concrete

Over 100 years, reinforced concrete has been extensively used as a traditional mixture of steel and concrete [93]. Steel provides tensile strength, while concrete prevents the steel from corrosion as a high alkalinity substance and as a physical barrier. The concrete [94] supplies the environment for the development of a passive layer that protects the steel reinforcement from corrosion with approximately 12-14 pH values [95]. This protection, though it is not adequate, since the substance has fissures and is porous in nature, enabling the diffusion of agents like chloride ions contributing to the reinforcement corrosion. The carbon dioxide also reacts with the hydrated cement products (which is identified as carbonation), results from the development of calcium carbonate layer, and thus decreases the pH, which destabilizes the passive reinforcement film and making it prone to corrosion [96]. Steel reinforcement corrosion is the major cause of reinforced concrete premature deterioration, resulting in large economic losses [97]. Chloride ions in a marine zone or the utilization of thaw salts [98] or carbonation in urban areas may cause brisk deterioration [99]. In recent years, many products recognized as CIs have been used to offer extra security and extend the life of reinforced structures [100]. By using CIs, the strength and carbonation of concrete structures exposed to chloride ions can be improved more efficiently [101]. CIs can be categorized as per their methods of applications, composition, or mechanism of protection [102]. These CIs can be organic and inorganic, which inhibits the process of corrosion. Basically, anodic inhibitors inhibit corrosion by forming a passive layer, whereas, with a cathodic inhibitor, it is achieved by decreasing corrosion potential and by enhancing polarity. Both types of acting inhibitors, after pretreatment, can be directly applied to the concrete. The pretreatment includes immersion of the reinforced concrete in the solution of corrosion inhibitor, adding it to the kneading liquid during the mixing of concrete, or applying it to the reinforced concrete framework layer.

Depending upon the source or origin, the inhibitors are also categorized as natural or synthetic. Synthetic inhibitors can cause human and environmental harm, particularly synthetic organic inhibitors. Many studies have explored the usefulness and viability of using green inhibitors, which are eco-friendly and sustainable because of the aforesaid problems [103].

Red mud is a residue produced in the processing of aluminum bauxite and included in the classification of green inhibitors. The addition of red mud in different concentrations to cement mass content significantly decreases the rate of corrosion [104].

Recently, the ginger extract is used as an eco-friendly/green corrosion inhibitor in stimulated concrete pore (SCP) solutions to enhance the chloride-induced corrosion resistance of reinforced steel. Stereo-microscope and various electrochemical methods were used to test the inhibiting efficiency of ginger extract and three other inhibiting systems. Liquid chromatography-mass spectrometry (LC-MS) confirmed the main phenolic compounds of ginger extract. The ginger extract worked as a mixed type CI that enhanced carbon steel's (C-steel) corrosion-resistant quality, mainly by forming a carbonaceous organic film to prevent both cathodic and anodic corrosion [105].

The green CIS extracted from orange peel waste also showed better inhibition. The orange peel (dry form) was extracted at 6-hour immersion time in methanol at 60 mbar pressure and 40 °C. The electrochemical polarization measurements and weight-loss testing for rebar immersion time (7 days) demonstrated good inhibition in aqueous NaCl solutions of 3.5 wt%. The steel rebar showed 0.02 mm/year decrease in rebar corrosion rates at the inhibitor concentration of 3% [106]. *Azadirachta indica* (neem) powder and dehydrated *aloe-vera* were used as a CI for concrete (M25 grade) steel. With a salinity of about 3.5%, the concrete was immersed in a solution. It was found that the concrete using *Azadirachta indica* (neem) powder as green CI has reduced the corrosion rate of steel on concrete more efficiently i.e., 0.22 mm/y compared to *aloe-vera* i.e., 0.27 mm/y [107]. Similarly, the leaf extract of *azadirachta indica* (neem) and *ruta graveolens* were investigated as organic inhibitors and compared with inorganic inhibitors (NaNO$_3$ and disodium dehydrate salt of EDTA). During the mixing of the reinforcing steel bar and concrete (OPC 43 grade, 3.279 specific gravity), inhibitors were introduced. Rebar was immersed in different solutions (HCl, NaCl and MgSO$_4$) for corrosion testing. The experimental results in 5% HCl and 5% NaCl showed the most promising potential of *azadirachta indica* extracts followed by *ruta graveolens*, NaNO$_3$ and Disodium dehydrated EDTA. Whereas in the case of 5% MgSO$_4$, disodium dehydrated EDTA had the most positive potential for corrosion [108]. In other researches, the leaf extract of *Vernonia amygdalina* containing alkaloids, tannin and saponin, were used as a corrosion inhibitor for C-steel (40 grade) and compared with inorganic inhibitors i.e., NaNO$_3$ and Ca(NO$_3$)$_2$. Rebar was immersed in 3.5% solution of NaCl for 56 days to investigate corrosion inhibition efficiency. As per weight loss analysis, NaNO$_3$ showed the highest inhibition efficiency, i.e., 96%. The IE of Ca(NO$_3$)$_2$ was 91% for 2% v/v inhibitor, whereas *Vernonia amygdalina*, had a 75% inhibition output with 6% v/v [109]. Another plant extract of *Prosopic juliflora* was investigated for corrosion IE for steel in concrete. Rebar was immersed in 3.5% solution of NaCl for 30 days. The corrosion test was performed with and without the inhibitor. It was discovered from the corrosion test that at the concentration of 120 ppm, the corrosion inhibitor would achieve the efficiency of 91%, but the efficiency found to be less (51%) at low inhibitor concentration i.e., at 100 ppm. According to electrochemical impedance spectroscopy (EIS) studies, there was a process of diffusion and precipitation effect of the solid $Ca(OH)_2$ layer at the interface of steel/concrete to form a protective layer. In addition, the increasing inhibitor concentration would minimize the inhomogeneity of the surface due to the adsorption of the extract molecules over the embedded steel layer. It was seen from AFM images that inhibitor molecules were absorbed on the rebar surface resulting in the formation of a layer that reduced the rebar's corrosion rate [110]. Green tea extract (the richest source of antioxidants), was compared to $CaNO_3$ corrosion inhibitor for their corrosion inhibition efficiency (IE) at the same volume on steel reinforcing bars embedded in a mortar. The IE of Green tea extract was significantly higher i.e., 75-80% due to a decrease in the iron oxidation rate and increased resistance to polarization potential. Green Tea extract was found to be a mixed-type inhibitor [111]. Green organic inhibitors form a barrier between a metal surface and a corrosive environment. In one research, the extract of *azadirachta indica* (neem) powder (organic inhibitors), dehydrated *aloe-vera* powder (organic inhibitors), and $Ca(NO_3)_2$ (inorganic inhibitors) were used to evaluate the corrosion IE of reinforced concrete in NaCl solution. The reinforcement corrosion resistance was tested using weight loss and accelerated corrosion analysis. The tests of the tested inhibitors confirmed that *Azadirachta indica* has a higher efficiency of corrosion inhibition compared to the *aloe-vera*. The chemical inhibitor of corrosion showed good inhibition of corrosion among all. There are stronger strength properties among all the chemical CIs. Yet green CIs have almost the same strength properties as traditional concrete at the same time. The adding up of green inhibitors does not affect the hardened properties of concrete significantly [112].

3.3   Acid pickling industry

Acid pickling has been widely used as an efficient method for cleaning metal surfaces, for example, by dipping steel into an acidic solution [113] to remove scale and rust from surfaces. Using acid to remove rust and debris from a metal surface is an effective means of pretreating metal surfaces and to remove scales from boilers and pipelines. The process of acid picking has been broadly utilized for the maintenance of machine materials, chemical industry, metallurgy, and many other industries [114]. Due to the corrosive nature of acids, in the acid pickling process, and overarching phenomenon occurs [115]. As a result, the acid pickling process not only dissolve the metal and waste the acid pickling solution but also generates a huge amount of acid pickling waste and ultimately causes deterioration of the environment. Generally, a CI is added to the acid solution at a low concentration to suppress the dissolution of the metal. In case, if the inhibitor is highly toxic, it will lead to secondary pollution [116]. An environmentally friendly CI is, therefore, a critical requirement for acid picking and cleaning. Some conventional CIs are restricted because they cause damage and pollute the environment [117]. However, plant-based corrosion inhibitors are non-toxic and eco-friendly; therefore, be steadily developed and extensively used in the pickling industry [118]. In the aggressive media, most plant extracts were used against material deterioration such as: *Urtica dioica* [119], *Zanthoxylum alatum* [120], *Phyllanthus* [121] etc. Furthermore, ethanol, methanol or solvent in the aqueous form are used to extract the green CIs. The leaf extract of *Sesbania grandiflora* obtained by methanol extraction illustrated outstanding corrosion IE of 98% in HCl medium at 10,000 ppm [122]. The extract of *Schinopsis lorentzii,* obtained by water extraction, showed moderate IE of 66% at 2,000 ppm [123]. Another green CI of *Ruta chalepensis* demonstrated excellent IE of 99.17% in HCl medium against the hydrogen embrittlement. The substantial inhibition of 99.17% explained their significance in different industries [124].

Many researchers reported the superior corrosion inhibition nature of flavonoids (in lotus leaves) as these are absorbed more strongly on the metal surface [125]. *Solanum lasiocarpum L.* fruit is very rich in flavonoids [126,127] and can, therefore, be a CI. In addition, it has a broad source, and production is abundant. The *Solanum lasiocarpum L.* extract contained N, O, and S-based polar groups. To determine the inhibition performance of the *Solanum lasiocarpum L.* extract, substances containing these polar groups could be used. With this aim, *Solanum lasiocarpum L.* extract were tested in a 1M HCl solution for its potential corrosion effect on A3-steel. Different techniques i.e., Electrochemical impedance spectroscopy (EIS), Fourier transform infrared spectroscopy (FTIR), weight-loss method, potentiodynamic polarization, and scanning electron microscopy (SEM) were utilized for the testing. The results of the SEM, electrochemical, and weight loss confirmed *Solanum lasiocarpum L.* extract as an effective CI. The *Solanum lasiocarpum L.* extract was a mixed inhibitor that suppresses the cathodic reaction predominantly, and at concentration 1 g/l, the IE was up to 93.31% [128].

Recently, a green CI, *Senecio anteuphobium* extract, was obtained by ethanolic extraction. Potentiodynamic polarization and EIS were employed to evaluate this inhibitory effect of *Senecio anteuphobium* extract on S300 steel in 1M HCl. With an increase in extract concentration up to 30 mg/L, the IE also increased to 91% [129].

Camellia sinensis extract is an environmental friendly and cost-effective natural CI. *Camellia sinensis* extract was investigated in 1 M HCl solution as a green CI for mild steel. The techniques used for investigation were potentiodynamic polarization, weight-loss method, and surface response methodology (RSM) modeling. The IE increased with increasing temperature, concentration and the time of immersion. At a concentration 800 ppm, *Camellia sinensis* extract demonstrated maximum IE, i.e., 90% at 353 k temperature with an immersion time of 5 hours [130].

3.4   Coatings

The extensively used commercial methods to protect metallic substrates from corrosion are the application of surface coatings [131]. Due to the highly toxic and carcinogenic limitations, researches on green substitutes of chromate-based coatings have been extensively investigated in recent years [132]. Various researches proved composite coatings as an effective corrosion barrier [133]. The barrier properties may slow down the rate of diffusion of the electrolyte to the metal substratum. But, the existence of local damages to the surface of the coating can make it easier for the aggressive electrolyte to penetrate the interface of the metal, thus severely restraining its stability [134]. The consistent and efficient approach for long term sustainable coatings is achieved by adding green CIs [135-142].

Zeolite coatings with anti-corrosion property are an emerging concern that can propose green substitute with superior performance to traditional protective coatings based on chromate. Because of the porosity of the zeolite structure, these coatings provide many versatile functions (high surface area, the storage capacity of ion and molecule, pore-volume, and molecular selectivity), resulting from its application in various industries. In this perspective, much research efforts have concentrated on anti-corrosion efficiency and safety of composite zeolite coatings, suggesting various deposition methods and forms of a matrix. The technology associated with the coatings of composite zeolite has been categorized into two i.e., sol-gel and thermosetting, showing such methods as the most advanced and appropriate for their applicability in the industrial sector.

Sol-gel technology is versatile and cost-effective that can meet the consumer's requirements [143]. The long aging and durability of coating are the limits of the composite sol-gel coatings, more research on the interaction of coating with the substrate as well as kinetic of inhibitor release are needed over time [144]. The coatings based on composite zeolite do need more improvement to address the various industrial sector-induced application challenges. In addition to the above-mentioned long-lasting characteristics, other characteristics like hydrophobicity, aesthetics, and environmental sustainability need to be considered [145-149]. More success in evaluating the development of defect-free zeolite films to avoid problems [150] is achieved by selecting perfect zeolite filler for individual composite to maximize a similar zeolite framework as a reference for the specific application [151]. The coatings that can provide more function (electrical conductivity, biocompatibility, etc.) in combination with corrosion resistance are the upcoming technology [152-155].

Graphene, a nanofiller with a single-atom-thick planar carbon layer (sp^2-bonded C-atoms), provides efficient corrosion resistance to the coatings. Several approaches of graphite sheet preparation have been studied, including chemical graphite exfoliation, mechanical graphite cleavage, direct synthesis, and thermal-induced exfoliation. The IE of copper (Cu) and nickel (Ni) were evaluated after developing graphene on their surface. Any of the two methods i.e., chemical vapor deposition (CVD) or mechanically transferring, can be used for developing multilayer graphene. Graphene, developed by the technique of CVD, showed higher anti-corrosion property. However, it was not practical to use this coating for a long time. Transferring multiple layers of graphene to the metal surfaces documented to improve protection to building thick and more durable films [156].

Strong thermal conductivity, enhanced gas barriers, high strength, exceptional properties of electronic transport coupled with a broad range of various peculiar properties allowed graphene composites as a competitive and cheaper alternative to composites based on carbon nanotubes [157-159]. Graphene and its derivatives can be utilized in sensors [160], batteries [161,162], storage of hydrogen [163], solar cells [164], capacitors [165], coatings of nanocomposites [166-168] and conductive films (transparent) [169,170]. Polyaniline/graphene composites in aqueous NaCl solution were used for steel corrosion inhibition. Exceptional inhibition potential was displayed by the composites against reactive gases (O_2 and H_2O). The diffusion path for O_2 and H_2O gases in the polymer was improved by using graphene in the coating matrix, resulting in an outstanding enhancement in corrosion inhibition in comparison to the normal coating of the polymer [171]. In another study, graphene sheets were incorporated in epoxy polymer composite resulting in decreased thermal expansion coefficients and better thermal conductivity [172]. The graphene can be coated on microelectronics due to its high thermal stability. As a new area of corrosion prevention technology, functionalized graphene will act as conductive nanofiller for other polymer/graphene-based composite coatings.

Stronger interaction with graphene platelets was achieved by utilizing multi-layer planes. The nanocomposites of epoxy/graphite were developed by adding epoxy and graphite in the solvent. By adding graphite (4 wt%), a marginal increase was observed in the glass transition temperature, whereas Young's module increases by 10%. Improved results (15% modulus) were achieved by using sonication and shear mixing 1 wt% graphene

Theory and Applications of Green Corrosion Inhibitors   Materials Research Forum LLC
Materials Research Foundations **86** (2021) 127-160   https://doi.org/10.21741/9781644901052-5

platelets, however, resulting in reduced tensile strength [173]. The mechanical properties also enhanced in epoxy/ graphene pellets nanocomposites. The toughness (0.1 wt% filler fraction) increased from 0.97 MPa m$^{1/2}$ to 1.48 MPa m$^{1/2}$ [174], and in coatings, also it provides toughness.

Although the ability of graphene nano-sheets to re-aggregate and stack, their use in polymer nanocomposites has been limited due to their strong Van der Waals forces and high surface area. Though, various researchers have paid attention to improve graphene interfaces and dispersion in a polymer matrix using functionalized graphene. A new approach to develop functionalized graphene was achieved by coupling agents in order to form a covalent bond between a soft matrix and fillers. The functionalization at 1 wt% fraction of filler enhanced the modulus by 50% [175]. With the same method, the fire resistance property of epoxy/graphene pellet nanocomposites was improved [176]. The functionalized graphene sheets/epoxy coatings were synthesized and evaluated on the mechanical properties of the coating. Results indicated higher stiffness and higher glass transition temperature with superior storage module properties [177]. The other film of grapheme/epoxy composite was synthesized by utilizing solution casting, which followed thermal curving. The films can be used in various applications as high-performance electrical heating elements. The important factors in regulating the maximum temperature of composite films were the graphene material, and the voltage applied [178]. The functionalized graphene oxides/epoxy nanocomposites improved electrical, thermal, and mechanical properties. The functionalized graphene oxide was prepared by utilizing in situ thermal polymerization [179].

Graphene is advantageous because of its anti-corrosion properties; it provides protection to the metal and preserves the intrinsic properties, which cannot be accomplished with 3D protective paints, polymers, or oxides. The biggest hurdle of using graphene as a corrosion-resistant substance is that there are still too many defects in the graphene sheets synthesized using current methods. Therefore, in the future, the major challenge is to develop a better quality of the sheet made as to the low-quality sheets significantly reduce the efficiency of corrosion inhibition. Missing of bonds, variations in local thickness, lattice distortion, presence of impurities are some main factors that affect the quality of graphene sheets. All these variables can also reflect the source of accumulation of damage other than altering the graphene's properties. Currently, the functionalization of grapheme oxides through covalent and non-covalent using organic compounds has become an emerging topic of research with highly developed functions for the development of revolutionary hybrid nanocomposites.

3.5 Aircraft industry

Most components of aircraft are made of metal and are highly prone to oxidation, like any other metal object. Since the Wright brothers built their first plane, Al has been used in the aerospace industry. Al and its alloys are important materials for structural engineering. Aerospace today, like many industries, makes wide use of the manufacture of Al. Because of its valuable properties, Al has become ideal for reducing aircraft manufacturing costs while maintaining the industry's stringent specifications. The main reason for extensive use of Al is due to its toughness, strength ratio: density, and to some extent, its resistance to corrosion [180,181]. On atmospheric exposure, an invisible thin protective oxide layer develops instantly on the Al surface. This self-protective characteristic offers high corrosion resistance to aluminum. Unless Al is exposed to some material or condition that removes the protective oxide layer, it remains entirely protected from corrosion. Alkalis and some acids are amongst the few compounds that strike the surface of the oxide film and are, therefore, Al corrosive [182].

The planes/aircraft usually spend many hours or days at the terminal, making the situation even worse as airborne salinity/acidity is likely to be collected on the body of aircraft. In reality, the phenomenon of corrosion is an omnipresent and inevitable risk to airworthiness and aircraft security. In addition to the external environment, it may be caused by trapped moisture that may enter the aircraft by a rainstorm or may result from rapid temperature fluctuations, particularly in tropical climates [183]. As a result, some corrosion modes commonly observed in aircraft are uniform corrosion, exfoliation corrosion, crevice corrosion, hydrogen embrittlement, pitting corrosion, galvanic corrosion, and stress corrosion [184]. The corrosion can be prevented by the selection of proper metal i.e., Al and its alloys, with the support of inhibitors in the presence of certain mild alkalis and acids. Direct contact can usually be prevented. As already discussed, the inorganic inhibitors are toxic, so green organic CIs are used to suppress the rate of corrosion [185].

The leaves extract of Areca Palm was used to test IE for Al in 0.5 M HCl by different techniques. The protection rates of Al enhanced with increased concentration of extract and decreased with an increase in contact time and temperature. According to the results of potentiodynamic polarization, gravitational tests, and SEM, it acted as a potential green CI with an efficiency of 89.493%, 88.114% and 97.035%, respectively [186]. In other studies, the leaf extract of *Stachytarpheta indica* was utilized to study the corrosion rate of Al in acidic solutions at different temperatures. The results of the thermometric and weight loss method revealed that at 30 °C and 60 °C, different concentrations of extract were active against the Al corrosion. The IE was observed to decrease with increasing extract concentration but improved with elevated temperature. The chemical adsorption of the extract on the surface of the Al was indicated by thermodynamic parameters [187]. Similarly, the root extract of *Maesobatrya barteri* reported as an efficient green CI for Al in acidic solutions [188]. While in an alkaline medium (1M NaOH), the seed extract of *Piper longum* was tested as a green CI for Al. *Piper longum* seed extract was an effective mixed-type corrosion inhibitor and suppressed both kinds of reactions (anodic and cathodic) on the Al surface. The IE increased on increasing the concentration of the *Piper longum* seed extract. The inhibition performance was significantly increased to 94% at an extract concentration of 400 mg/L [189].

The corrosion rate of AA8011 alloy of Al in alkaline medium (0.25M KOH) was achieved by using *Palisota hirsute*, an environment-friendly inhibitor. *Palisota hirsute* decreased the corrosion rate from 0.407 mm/yr to 0.106 mm/yr after exposing for 5 hrs. The IE increased on increasing the concentration of the extract. Through the physical adsorption mechanism, the inhibitor molecules were spontaneously adsorbed to the Al surface [190]. Another green CI i.e., the aqueous bark extract of *Neolamarkia Cadamba* was investigated against Al alloy in alkaline medium (1M NaOH). Various techniques, such as weight-loss method, polarization resistance, EIS, and Tafel polarization, were employed for characterization. *Neolamarkia Cadamba* bark extract acted as a mixed-type inhibitor and formed a protective film on the Al alloy surface. Here also the same trend was observed i.e., with increased concentration (0.6 g/L), the IE was increased (87%) [191]. There are various different extract of plants that provides the inhibition to the Al metal, such as, *Jasminum nudiflorum Lindl* [192], *Morinda tinctoria* [193], *Dendrocalamus brandisii* [194], *Bacopamonnieri* [195], *Morus nigra L.* [196], *Laurus nobilis L.* [197], *Aspilia Africana* [198], *Trachyspermum copticum* [199], *Newbouldia leavis* [200,201] and numerous other plant extracts have been reported to have IE.

3.6   Water industry

Rain and snow are the sources of fresh water and consumed by humans as potable water. Fresh water in rivers and lakes contain chlorides, phosphates, and sulfates (Total dissolved solids). Human consumes freshwaters as potable water which is conveyed *via* pipelines. These water pipelines are generally made up of C-steels and severely affect the quality of water if it suffers any type of corrosion. The water industry utilizes inhibitors to avoid corrosion. A number of green corrosion inhibitors are employed to reduce the degradation of material (C-steel) used in power stations, wastewater plants, desalination plants, potabilization plants. The Olive Leaf in 10% NH_2SO_3H solution acted as a mixed-type CI for C-steel and tested using various techniques. The percentage of IE increased as the concentration of olive leaf extract increased. Using Olive Leaf resulted in reduced double-layer capacitance and enhanced charge transfer resistance [202]. Similarly, various plant extracts were evaluated as efficient green inhibitors against C-steel [203-205].

Conclusion

Plant extracts are suitable for replacing conventional inorganic and man-made organic CIs that are costly and poisonous. These extracts consist of various components that can easily be absorbed by metallic corrosion and inhibit it. Diverse forms of plant extracts like leaves, stems, roots, etc. in different environments (acidic, basic, and neutral) used as efficient and active green CIs for C-steel, mild steel, Al metals, and alloys. Green CIs have tremendous technological applications and are broadly utilized for corrosion control in industrial plants like reinforced concrete, coating, aircraft, oil and gas, acid pickling, and water industry.

Acknowledgment

Dr. Pallavi Jain is grateful to Dean and Deputy Registrar of SRM Institute of Science & Technology, Modinagar, for the encouragement and for providing the facilities to research.

References

[1] P. Singh, V. Srivastava, M.A. Quraishi, Novel quinoline derivatives as green corrosion inhibitors for mild steel in acidic medium: Electrochemical, SEM, AFM, and XPS studies, J. Mol. Liq. 216 (2016) 164-173. https://doi.org/10.1016/j.molliq.2015.12.086

[2] N. Kicir, G. Tansug, M. Erbil, T. Tuken, Investigation of ammonium (2, 4-dimethylphenyl)-dithiocarbamate as a new, effective corrosion inhibitor for mild steel, Corros. Sci. 105 (2016) 88-99. https://doi.org/10.1016/j.corsci.2016.01.006

[3] L.Li, X. Zhang, S. Gong, H. Zhao, Y. Bai, Q. Li, L. Ji, The discussion of descriptors for the QSAR model and molecular dynamics simulation of benzimidazole derivatives as corrosion inhibitors, Corros. Sci. 99 (2015) 76-88. https://doi.org/10.1016/j.corsci.2015.06.003

[4] H.Q. Yang, Q. Zhang, S.S. Tu, Y. Wang, Y.M. Li, Y. Huang, Effects of inhomogeneouselastic stress on corrosion behaviour of Q235 steel in 3.5% NaCl solution using a novel multichannel electrode technique, Corros. Sci. 110 (2016) 1-14. https://doi.org/10.1016/j.corsci.2016.04.017

[5] W. Mai, S. Soghrati, R.G. Buchheit, A phase field model for simulating the pitting corrosion, Corros. Sci. 110 (2016) 157-166. https://doi.org/10.1016/j.corsci.2016.04.001

[6] R. Hummel, Alternative futures for corrosion and degradation research, Potomac Institute Press, Arlington, 2014.

[7] R. Mohammadinejad, S. Karimi, S. Iravani, R.S. Varma, Plant-derived nanostructures: types and applications, Green Chem. 18 (2016) 20-52. https://doi.org/10.1039/C5GC01403D

[8] A. Singh, S. Mohapatra, B. Pani, Corrosion inhibition effect of Aloe Vera gel: Gravimetric and electrochemical study, J. Ind. Eng. Chem. 33 (2016) 288-297. https://doi.org/10.1016/j.jiec.2015.10.014

[9] M. Prabakaran, S.H. Kim, V. Hemapriya, M. Gopiraman, I.S. Kim, I.M. Chung, *Rhus verniciflua*as a green corrosion inhibitor for mild steel in 1 M H_2SO_4, RSC Adv. 6 (2016) 57144-57153. https://doi.org/10.1039/C6RA09637A

[10]G. Bahlakeh, M. Ramezanzadeh, B. Ramezanzadeh, Experimental and theoretical studies of the synergistic inhibition effects between the plant leaves extract (PLE) and zinc salt (ZS) in corrosion control of carbon steel in chloride solution, J. Mol. Liq. 248 (2017) 854-870. https://doi.org/10.1016/j.molliq.2017.10.120

[11]J. Haque, V. Srivastava, C. Verma, M.A. Quraishi, Experimental and quantum chemicalanalysis of 2-amino-3-((4-((S)-2-amino-2-carboxyethyl)-1H-imidazol-2-yl)thio)propionicacidas new and green corrosion inhibitor for mild steel in 1 M hydrochloric acid solution, J. Mol. Liq. 225 (2017) 848-855. https://doi.org/10.1016/j.molliq.2016.11.011

[12]S.R. Kumar, I. Danaee, M. RashvandAvei, M. Vijayan, Quantum chemical andexperimental investigations on equipotent effects of (+) R and (−) S enantiomers of racemic amisulpride as eco-friendly corrosion inhibitors for mild steel in acidic solution, J. Mol. Liq. 212 (2015) 168-180. https://doi.org/10.1016/j.molliq.2015.09.001

[13]M. Srivastava, P. Tiwari, S.K. Srivastava, R. Prakash, G. Ji, Electrochemical investigation of Irbesartan drug molecules as an inhibitor of mild steel corrosion in 1 M HCl and 0.5 M H_2SO_4 solutions, J. Mol. Liq. 236 (2017) 184-197. https://doi.org/10.1016/j.molliq.2017.04.017

[14]K. Anupama, K. Ramya, A. Joseph, Electrochemical and computational aspects of surface interaction and corrosion inhibition of mild steel in hydrochloric acid by

Phyllanthus amarus leaf extract (PAE), J. Mol. Liq. 216 (2016) 146-155.
https://doi.org/10.1016/j.molliq.2016.01.019

[15]M. ur Rahman, S. Gul, M. Umair, A. Anwar, A. Achakzai, Anticorrosive Activity of *Rosemarinu officinalis* L. Leaves Extract Against Mild Steel in Dilute Hydrochloric Acid, Inter. J. Innovative Res. Adv. Eng. 3 (2016) 385-43.

[16]C. Verma, E.E. Ebenso, M.A. Quraishi, Corrosion inhibitors for ferrous and non-ferrous metals and alloys in ionic sodium chloride solutions: A review, J. Mol. Liq. 248 (2017) 927-942. https://doi.org/10.1016/j.molliq.2017.10.094

[17]P. Singh, M. Makowska-Janusik, P. Slovensky, M.A. Quraishi, Nicotinonitriles as green corrosion inhibitors for mild steel in hydrochloric acid: Electrochemical, computational andsurface morphological studies, J. Mol. Liq. 220 (2016) 71-81. https://doi.org/10.1016/j.molliq.2016.04.042

[18]Y. Li, D. Wang,L. Zhang, Experimental and theoretical research on a new corrosion inhibitor for effective oil and gas acidification, RSC Adv. 9 (2019) 26464-26475. https://doi.org/10.1039/C9RA04638K

[19]M. Basik, M. Mobin, M. Shoeb, Cysteine-silver-gold Nanocomposite as potential stable green corrosion inhibitor for mild steel under acidic condition, Sci. Rep. 10 (2020) 279. https://doi.org/10.1038/s41598-019-57181-5

[20]P. Dauthal, M. Mukhopadhyay, Noble metal nanoparticles: plant-mediated synthesis, mechanistic aspects of synthesis, and applications, Ind. Eng. Chem. Res. 55 (2016) 9557-9577. https://doi.org/10.1021/acs.iecr.6b00861

[21]B. Sanyal, Organic compounds as corrosion inhibitors in different environments-A review, Prog. Org. Coat. 9 (1981) 165-236. https://doi.org/10.1016/0033-0655(81)80009-X

[22]K. Hu, J. Zhuang, C. Zheng, Z. Ma, L. Yan, H. Gu, X. Zeng, J. Ding, Effect of novelcytosine-l-alanine derivative based corrosion inhibitor on steel surface in acidic solution, J. Mol. Liq. 222 (2016) 109-117. https://doi.org/10.1016/j.molliq.2016.07.008

[23]B. Ramezanzadeh, H. Vakili, R. Amini, The effects of addition of poly (vinyl) alcohol(PVA) as a green corrosion inhibitor to the phosphate conversion coating on the anticorrosion and adhesion properties of the epoxy coating on the steel substrate, Appl. Surf. Sci. 327(2015) 174-181. https://doi.org/10.1016/j.apsusc.2014.11.167

[24] P. Parthipan, J. Narenkumar, P. Elumalai, P.S. Preethi, A.U.R. Nanthini, A. Agrawal, A. Rajasekar, Neem extract as a green inhibitor for microbiologically influenced corrosion of carbon steel API 5LX in a hypersaline environments, J. Mol. Liq. 240 (2017) 121-127. https://doi.org/10.1016/j.molliq.2017.05.059

[25] E. Alibakhshi, M. Ramezanzadeh, G. Bahlakeh, B. Ramezanzadeh, M. Mahdavian, M. Motamedi, *Glycyrrhiza glabra* leaves extract as a green corrosion inhibitor for mild steel in 1 M hydrochloric acid solution: Experimental, molecular dynamics, Monte Carlo and quantum mechanics study, J. Mol. Liq. 255 (2018) 185-198. https://doi.org/10.1016/j.molliq.2018.01.144

[26] S. Mo, L.J. Li, H.Q. Luo, N.B. Li, An example of green copper corrosion inhibitors derivedfrom flavor and medicine: Vanillin and isoniazid, J. Mol. Liq. 242 (2017) 822-830. https://doi.org/10.1016/j.molliq.2017.07.081

[27] M.V. Diamanti, U.V. Velardi, A. Brenna, A. Mele, M. Pedeferri, M. Ormellese, Compatibility of imidazolium-based ionic liquids for CO_2 capture with steel alloys: a corrosion perspective, Electrochim. Acta 192 (2016) 414-421. https://doi.org/10.1016/j.electacta.2016.02.003

[28] S. Yesudass, L.O. Olasunkanmi, I. Bahadur, M.M. Kabanda, I. Obot, E.E. Ebenso, Experimental and theoretical studies on some selected ionic liquids with different cations/anions as corrosion inhibitors for mild steel in acidic medium, J. Taiwan Inst. Chem. E 64 (2016) 252-268. https://doi.org/10.1016/j.jtice.2016.04.006

[29] V. Srivastava, J. Haque, C. Verma, P. Singh, H. Lgaz, R. Salghi, M.A. Quraishi, Amino acid based imidazolium zwitterions as novel and green corrosion inhibitors for mild steel: Experimental, DFT and MD studies, J. Mol. Liq. 244 (2017) 340-352. https://doi.org/10.1016/j.molliq.2017.08.049

[30] F. El-Hajjaji, M. Messali, A. Aljuhani, M. Aouad, B. Hammouti, M. Belghiti, D. Chauhan, M.A. Quraishi, Pyridazinium-based ionic liquids as novel and green corrosion inhibitors of carbonsteel in acid medium: Electrochemical and molecular dynamics simulation studies, J. Mol. Liq. 249 (2018) 997-1008. https://doi.org/10.1016/j.molliq.2017.11.111

[31] S.K. Shetty, A.N. Shetty, Eco-friendly benzimidazolium based ionic liquid as a corrosion inhibitor for aluminum alloy composite in acidic media, J. Mol. Liq. 225 (2017) 426-438. https://doi.org/10.1016/j.molliq.2016.11.037

[32] R.A. Sheldon, Green chemistry and resource efficiency: towards a green economy, Green Chem. 18 (2016) 3180-3183. https://doi.org/10.1039/C6GC90040B

Materials Research Forum LLC
https://doi.org/10.21741/9781644901052-5

[33] M. Chellouli, D. Chebabe, A. Dermaj, H. Erramli, N. Bettach, N. Hajjaji, M. Casaletto, C. Cirrincione, A. Privitera, A. Srhiri, Corrosion inhibition of iron in acidic solution by a green formulation derived from *Nigella sativa* L, Electrochim. Acta 204 (2016) 50-59. https://doi.org/10.1016/j.electacta.2016.04.015

[34] N. M'hiri, D. Veys-Renaux, E. Rocca, I. Ioannou, N.M. Boudhrioua, M. Ghoul, Corrosion inhibition of carbon steel in acidic medium by orange peel extract and its main antioxidant compounds, Corros. Sci. 102 (2016) 55-62. https://doi.org/10.1016/j.corsci.2015.09.017

[35] S.A. Umoren, M.M. Solomon, Synergistic corrosion inhibition effect of metal cations and mixtures of organic compounds: A review, J. Environ. Chem. Eng. 5 (2017) 246-273. https://doi.org/10.1016/j.jece.2016.12.001

[36] C. Verma, E.E. Ebenso, M.A. Quraishi, Ionic liquids as green and sustainable corrosion inhibitors for metals and alloys: An overview, J. Mol. Liq. 233 (2017) 403-414. https://doi.org/10.1016/j.molliq.2017.02.111

[37] K. Rose, B.S. Kim, K. Rajagopal, S. Arumugam, K. Devarayan, Surface protection of steel in acid medium by *Tabernaemontana divaricata* extract: Physicochemical evidence for adsorption of inhibitor, J. Mol. Liq. 214 (2016) 111-116. https://doi.org/10.1016/j.molliq.2015.12.008

[38] B. Sanyal, Organic compounds as corrosion inhibitors in different environments-A review, Prog. Org. Coat. 9 (1981) 165-236. https://doi.org/10.1016/0033-0655(81)80009-X

[39] R.S. Varma, Greener and sustainable trends in synthesis of organics and nanomaterials, ACS Publications, ACS Sustainable Chem. Eng. 4 (2016) 5866-5878. https://doi.org/10.1021/acssuschemeng.6b01623

[40] M. Jokar, T.S. Farahani, B. Ramezanzadeh, Electrochemical and surface characterizations of *Morus Alba Pendula* leaves extract (MAPLE) as a green corrosion inhibitor for steel in 1 M HCl, J. Taiwan Inst. Chem. Eng. 63 (2016) 436-452. https://doi.org/10.1016/j.jtice.2016.02.027

[41] S. Kuppusamy, T. Palanisami, M. Megharaj, K. Venkateswarlu, R. Naidu, In-situ remediation approaches for the management of contaminated sites: a comprehensive overview, Rev. Environ.Contam.T236 (2016) 1-115. https://doi.org/10.1007/978-3-319-20013-2_1

[42] K. Anupama, K. Shainy, A. Joseph, Excellent anticorrosion behavior of *Ruta Graveolens* extract (RGE) for mild steel in hydrochloric acid: electro analytical

studies on the effect of time, temperature, and inhibitor concentration, J. Bio-and Tribo-Corros. 2 (2016) 1-10. https://doi.org/10.1007/s40735-016-0032-5

[43] P. Deivanayagam, I. Malarvizhi, S. Selvaraj, Alcoholic extract of *"Gymnema Sylvestre"* leaves on mild steel in acid medium Quinquefasciatus say, Int. J. Adv. Sci. Res. 1 (2016) 21-27.

[44] Y. Chen, Y. Zhou, Q. Yao, Y. Bu, H. Wang, W. Wu, W. Sun, Evaluation of a low phosphorus terpolymer as calcium scales inhibitor in cooling water, Desalin. Water Treat. 55 (2015) 945-955. https://doi.org/10.1080/19443994.2014.922500

[45] S. Firdausi, F. Kurniawan, Corrosion inhibition by *Tithonia diversifolia* (Hemsl) A. Gray leaves extract for 304 SS in hydrochloric acid solution, J. Phys.: Conf. Ser., IOP Publishing, 710 (2016) 012042. https://doi.org/10.1088/1742-6596/710/1/012042

[46] K. Ramya, N. Muralimohan, Study on corrosion inhibitor in mild steel by various habitual plant extract-review, Int. J. Chemical Concepts 2 (2016) 70-75.

[47] K.H. Hassan, A.A. Khadom, N.H. Kurshed, *Citrus aurantium* leaves extracts as a sustainable corrosion inhibitor of mild steel in sulfuric acid, S. Afr. J. Chem. Eng. 22 (2016) 1-5. https://doi.org/10.1016/j.sajce.2016.07.002

[48] A. Singh, *Cassia tora* leaves extract as mild steel corrosion inhibitor in sulphuric acid solution, Imperial J. Interdisciplinary Res. 2 (2016) 698-701.

[49] S. Umoren, U. Eduok, M. Solomon, A. Udoh, Corrosion inhibition by leaves and stem extracts of Sidaacuta for mild steel in 1 M H_2SO_4 solutions investigated by chemical and spectroscopic techniques, Arab. J. Chem. 9 (2016) S209-S224. https://doi.org/10.1016/j.arabjc.2011.03.008

[50] O.O. Dominic, O. Monday, Optimization of the inhibition efficiency of *Mango* extract as corrosion inhibitor of mild steel in 1.0M H_2SO_4 using response surface methodology, J. Chem. Technol. Metall. 51 (2016) 302-314.

[51] H. Wang, M. Gao, Y. Guo, Y. Yang, R. Hu, A natural extract of tobacco rob as scale and corrosion inhibitor in artificial seawater, Desalination 398 (2016) 198-207. https://doi.org/10.1016/j.desal.2016.07.035

[52] M.J. Meften, N.Z. Rajab, M.T. Finjan, Synthesis of new heterocyclic compound used as corrosion inhibitor for crude oil pipelines, Am. Sci. Res. J. Eng. Technol. Sci. 27 (2017) 419-437.

[53]O.U. Abakedi, V.F. Ekpo, E.E. John, Corrosion inhibition of mild steel *by* Stachytarpheta indica leaf extract in acid medium, Pharma. Chem. J. 3 (2016) 165-171.

[54]M. Omotioma, O. Onukwuli, Evaluation of pawpaw leaves extract as anti-corrosion agent for aluminium in hydrochloric acid medium, Niger. J. Tech. 36 (2017) 496-504. https://doi.org/10.4314/njt.v36i2.24

[55]M.S. Singh, S. Chowdhury, Recent developments in solvent-free multicomponent reactions: a perfect synergy for eco-compatible organic synthesis, RSC Adv. 2 (2012) 4547-4592. https://doi.org/10.1039/c2ra01056a

[56]Q. Deng, H.W. Shi, N.N. Ding, B.Q. Chen, X.P. He, G. Liu, Y. Tang, Y.T. Long, G.R. Chen, Novel triazolyl bis-amino acid derivatives readily synthesized via click chemistry as potential corrosion inhibitors for mild steel in HCl, Corros. Sci. 57 (2012) 220-227. https://doi.org/10.1016/j.corsci.2011.12.014

[57]A. Aljuhani, W.S. El-Sayed, P.K. Sahu, N. Rezki, M.R. Aouad, R. Salghi, M. Messali, Microwave-assisted synthesis of novel imidazolium, pyridinium and pyridazinium-based ionic liquids and/or salts and prediction of physico-chemical properties for their toxicity and antibacterial activity, J. Mol. Liq.249 (2018) 747-753. https://doi.org/10.1016/j.molliq.2017.11.108

[58]R.C. Cioc, E. Ruijter, R.V. Orru, Multicomponent reactions: advanced tools for sustainable organic synthesis, Green Chem. 16 (2014) 2958-2975. https://doi.org/10.1039/C4GC00013G

[59]D.I. Njoku, I. Ukaga, O.B. Ikenna, E.E. Oguzie, K.L. Oguzie, N. Ibisi, Natural products for materials protection: Corrosion protection of aluminium in hydrochloric acid by Kola nitida extract, J. Mol. Liq. 219 (2016) 417-424. https://doi.org/10.1016/j.molliq.2016.03.049

[60]M. Deyab, Inhibition activity of seaweed extract for mild carbon steel corrosion in saline formation water, Desalination 384 (2016) 60-67. https://doi.org/10.1016/j.desal.2016.02.001

[61]R. Pathak, P. Mishra, Drugs as corrosion inhibitors: A review, Int. J. Sci. Res. 5 (2016) 671-677. https://doi.org/10.21275/v5i4.NOV162623

[62]C. Verma, D. Chauhan, M.A. Quraishi, Drugs as environmentally benign corrosion inhibitors for ferrous and nonferrous materials in acid environment: An overview, J. Mater. Environ. Sci. 8 (2017) 4040-4051.

[63]S.H. Zaferani, M. Sharifi, D. Zaarei, M.R. Shishesaz, Application of eco-friendly products as corrosion inhibitors for metals in acid pickling processes-A review, J. Environ. Chem. Eng. 1 (2013) 652-657. https://doi.org/10.1016/j.jece.2013.09.019

[64]M. Finsgar, J. Jackson, Application of corrosion inhibitors for steels in acidic media for theoil and gas industry: A review, Corros. Sci. 86 (2014) 17-41. https://doi.org/10.1016/j.corsci.2014.04.044

[65]D. Winkler, M. Breedon, A. Hughes, F. Burden, A. Barnard, T. Harvey, I. Cole, Towards chromate-free corrosion inhibitors: structure-property models for organic alternatives, Green Chem. 16 (2014) 3349-3357. https://doi.org/10.1039/C3GC42540A

[66]D. Kesavan, M. Gopiraman, N. Sulochana, Green inhibitors for corrosion of metals: A review, Chem. Sci. Rev. Lett. 1 (2012) 1-8.

[67]S.A. Umoren, U.M. Eduok, Application of carbohydrate polymers as corrosion inhibitors for metal substrates in different media: A review, Carbohyd. Polym. 140 (2016) 314-341. https://doi.org/10.1016/j.carbpol.2015.12.038

[68]C. Verma, L. Olasunkanmi, E.E. Ebenso, M.A. Quraishi, Substituents effect on corrosion inhibition performance of organic compounds in aggressive ionic solutions: A review, J. Mol. Liq. 251 (2017) 100-118. https://doi.org/10.1016/j.molliq.2017.12.055

[69]M. Antonijevic, M. Petrovic, Copper corrosion inhibitors. A review, Int. J. Electrochem. Sci. 3 (2008) 1-28.

[70]T.L. Metroke, R.L. Parkhill, E.T. Knobbe, Passivation of metal alloys using sol-gel derived materials-A review, Prog. Org. Coat. 41 (2001) 233-238. https://doi.org/10.1016/S0300-9440(01)00134-5

[71]M. Prabakaran, S.-H. Kim, K. Kalaiselvi, V. Hemapriya, I.-M. Chung, Highly efficient Ligularia fischeri green extract for the protection against corrosion of mild steel in acidic medium: Electrochemical and spectroscopic investigations, J. Taiwan Inst. Chem. Eng. 59 (2016) 553-562. https://doi.org/10.1016/j.jtice.2015.08.023

[72]C. Achebe, A. Ilogebe, J. Chukwuneke, O. Azaka, D. Ugwuegbu, Mild steel corrosion inhibition in h_2so_4 using ethanol extract of vernonia amygdalina, Inter. J. Eng. Sci., 2 (2016) 50-55.

[73]O.K. Abiola, N.C. Oforka, E.E. Ebenso, N.M. Nwinuka, Eco-friendly corrosion inhibitors: The inhibitive action of *Delonix Regia* extract for the corrosion of

aluminium in acidic media, Anti-Corros. Method M54 (2007) 219-224. https://doi.org/10.1108/00035590710762357

[74] M.H. Hussain, A.A. Rahim, M.N.M. Ibrahim, N. Brosse, The capability of ultra filtrated alkaline and organosolv oil palm (Elaeis guineensis) fronds lignin as green corrosion inhibitor for mild steel in 0.5 M HCl solution, Measurement 78 (2016) 90-103. https://doi.org/10.1016/j.measurement.2015.10.007

[75] M.A. Quraishi, D.K. Yadav, I. Ahamad, Green approach to corrosion inhibition by black pepper extract in hydrochloric acid solution, Open Corros. J. 2 (2009) 56-60. https://doi.org/10.2174/1876503300902010056

[76] E.E. Ebenso, U.J. Ibok, U.J. Ekpe, S. Umoren, E. Jackson, O.K. Abiola, N.C. Oforka, S. Maritinez, Corrosion inhibition studies of some plant extracts on aluminium in acidic medium, Trans. SAEST 39 (2004) 117-123.

[77] H. Liu, Pipeline Engineering, Lewis Publishers, Boca Raton London, New York, Washington, D.C. 2003.

[78] L.T. Popoola, A.S. Grema, G.K. Latinwo, B. Gutti, A.S.Balogun, Corrosion Problems during Oil and Gas Production and its Mitigation, Intern. J. Indus. Chem. 35 (2013) 1-15. https://doi.org/10.1186/2228-5547-4-35

[79] C.J. Houghton, R.V. Westermark, North sea down hole corrosion: identifying the problem; implementing the solutions, J. Pet. Technol. 35 (1983) 239-246. https://doi.org/10.2118/11669-PA

[80] M. Bonis, J.-L. Crolet, Practical aspects of the influence of *in situ* pH on H_2S-induced cracking, Corros. Sci. 27 (1987) 1059-1070. https://doi.org/10.1016/0010-938X(87)90098-9

[81] O. Fatoba, R. Akid, Low cycle fatigue behaviour of API 5l x65 pipeline steel at room temperature, Proc. Eng.74 (2014) 279-286. https://doi.org/10.1016/j.proeng.2014.06.263

[82] F.O. Kolawole, S.K. Kolawole, J.O. Agunsoye, J.A. Adebisi, S.A. Bello, S.B. Hassan, Mitigation of corrosion problems in API 5l steel pipeline-A review, J. Mater. Environ. Sci. 9 (2018) 2397-2410.

[83] N.O. Eddy, S.A. Odoemelam, Inhibition of the corrosion of mild steel in H_2SO_4 by ethanol extract of Aloe vera, Pigm. Res. Techn. 38 (2009) 111-115. https://doi.org/10.1108/03699420910940617

[84] A. Bouyanzer, B. Hammouti, L. Majidi, Pennyroyal oil from Mentha pulegium as corrosion inhibitor for steel in 1 M HCl, Mater. Lett. 60 (2006) 2840-2843. https://doi.org/10.1016/j.matlet.2006.01.103

[85] F.S. Souza, A. Spinelli, Caffeic acid as a green corrosion inhibitor for mild steel, Corros. Sci. 5 (2009) 642-649. https://doi.org/10.1016/j.corsci.2008.12.013

[86] P. Rajeev, A.O. Surendranathan, C.S.N. Murthy, Corrosion mitigation of the oil well steels using organic inhibitors-A review, J. Mater. Environ. Sci. 3 (2012) 856-869.

[87] A.K. Satapathy, G. Gunasekaran, S.C. Sahoo, K. Amit, P.V. Rodrigues, Corrosion inhibition by Justicia gendarussa plant extract in hydrochloric acid solution, Corros. Sci. 51 (2009) 2848-2856. https://doi.org/10.1016/j.corsci.2009.08.016

[88] D.M.A. Njokua, K.I.O. Chidiebere, Corrosion inhibition of mild steel in hydrochloric acid solution by the leaf extract of Nicotiana tabacum, Adv. Mater. Corros. 1 (2013) 54-61.

[89] J.J. Bhawsar, Experimental and computational studies of Nicotiana tabacum leaves extract as green corrosion inhibitor for mild steel in acidic medium, Eng. J. 54 (2015) 769-775. https://doi.org/10.1016/j.aej.2015.03.022

[90] M.F. Azmi, J.W. Soedarsono, Study of corrosion resistance of pipeline API 5L X42 using green inhibitor bawang dayak (Eleutherine Americanna Merr.) in 1M HCl, IOP Conf. Series: Earth and Environ. Sci. 105 (2018) 012061. https://doi.org/10.1088/1755-1315/105/1/012061

[91] N.C. Ngobiri, E.E. Oguzie, Y. Li, L. Liu, N.C. Oforka, O. Akaranta, Eco-friendly corrosion inhibition of pipeline steel using brassica oleracea, Int. J. Corr. 2015 (2015) 1-9. https://doi.org/10.1155/2015/404139

[92] N. Muthukumar, S. Maruthamuthu, N. Palaniswamy, Green inhibitors for petroleum product pipelines, Electrochemistry 75 (2007) 50-53. https://doi.org/10.5796/electrochemistry.75.50

[93] H. Shan, J. Xu, Z. Wanga, L. Jiang, N. Xu, Electrochemical chloride removal in reinforced concrete structures: Improvement of effectiveness by simultaneous migration of silicate ion, Constr. Build. Mater. 127 (2016) 344-352. https://doi.org/10.1016/j.conbuildmat.2016.09.137

[94] Y.P. Asmara, T. Kurniawan, A.G.E. Sutjipto, J. Jafar, Application of plants extracts as green corrosion inhibitors for steel in concrete - A review, IJST3 (2018) 158-170. https://doi.org/10.17509/ijost.v3i2.12760

Materials Research Forum LLC
https://doi.org/10.21741/9781644901052-5

[95] V. Elfmarkova, P. Spiesz, H.J.H. Brouwers, Determination of the chloride diffusion coefficient in blended cement mortars, Cem. Concr. Res. 78 (2015) 190-199. https://doi.org/10.1016/j.cemconres.2015.06.014

[96] D. Zhang, Z. Ghouleh, Y. Shao, Review on carbonation curing of cement-based materials, Journal of CO_2 Utilization 21 (2017) 119-131. https://doi.org/10.1016/j.jcou.2017.07.003

[97] F.L. Fei, J. Hu, J.X. Wei, Q.J. Yu, Z.S. Chen, Corrosion performance of steel reinforcement in simulated concrete pore solutions in the presence of imidazoline quaternary ammonium salt corrosion inhibitor, Constr. Build. Mater.70 (2014) 43-53. https://doi.org/10.1016/j.conbuildmat.2014.07.082

[98] H.S. Lee, H.S. Ryu, W.J. Park, M.A. Ismail, Comparative study on corrosion protection of reinforcing steel by using amino alcohol and lithium nitrite inhibitors, Materials (Basel) 8 (2015) 251-269. https://doi.org/10.3390/ma8010251

[99] E. Redaelli, L. Bertolini, Electrochemical repair techniques in carbonated concrete. Part I: Electrochemical realkalisation, J. Appl. Electrochem. 41 (2011) 817-827. https://doi.org/10.1007/s10800-011-0301-4

[100] A.S. Abdulrahman, M. Ismail, M.S. Hussain, Corrosion inhibitors for steel reinforcement in concrete: A review, Sci. Res. Essays 6 (2011) 4152-4162. https://doi.org/10.5897/SRE11.1051

[101] J. Xu, L. Jiang, W. Wang, L. Tang, L. Cui, Effectiveness of inhibitors in increasing chloride threshold value for steel corrosion. Water Science and Engineering 6 (2013) 354-363.

[102] P.B. Raja, S. Ghoreishiamiri, M. Ismail, Natural corrosion inhibitors for steel reinforcement in concrete-A review, World Scientific Publishing Company WSPC 22 (2015) 68-75. https://doi.org/10.1142/S0218625X15500407

[103] A.S. Abdulrahman, M. Ismail, Green plant extract as a passivation-promoting inhibitor forreinforced concrete, Int. J. Eng. Sci. Technol. 3(2011) 6484-6490.

[104] D.V. Ribeiro, M.P.T. Cunha Deterioracao das estruturas de concretoarmado. in: D.V. Ribeiro (Ed.), Corrosaoemestruturas de concretoarmado: Teoria, Controle e Metodos de Analise, Rio de Janeiro, Elsevier, 2014, pp. 87-118.

[105] Y. Liu, Z. Song, W. Wang, L. Jiang, Y. Zhang, M. Guo, F. Song, N. Xu, Effect of ginger extract as green inhibitor on chloride-induced corrosion of carbon steel in

simulated concrete pore solutions, J. Clean. Prod. 214 (2019) 298-307.
https://doi.org/10.1016/j.jclepro.2018.12.299

[106] A.S.E.F. Abbas, T.I. Torok, Corrosion studies of steel rebar samples in neutral
sodium chloride solution also in the presence of a bio-based (green) inhibitor, Int. J.
Corros. Scale Inhib. 7 (2018) 38-47.

[107] C. Lisha, M. Rajalingam, G. Sunilaa, Corrosion resistance of reinforced concrete
with green corrosion, Int. J. Eng. Sci. Invent. Res. Develop. 3 (2017) 687-691.

[108] G.M. Akshatha, B.G.J. Kumar, H. Pushpa, Effect of corrosion inhibitors in
reinforced concrete, Int. J. Innov. Res. Sci. Eng. Technol. 4 (2015) 6794-6801.
https://doi.org/10.15680/IJIRSET.2015.0408013

[109] D.G. Eyu, H. Esah, C. Chukwuekezie, J. Idris, I. Mohammad, Effect of green
inhibitor on the corrosion behaviour of reinforced carbon steel in concrete, ARPN J.
Eng. Appl. Sci. 8 (2013) 326-332.

[110] S. Palanisamy, G. Maheswaran, C. Kamal, G. Venkatesh, Prosopisjuliflora-A
green corrosion inhibitor for reinforced steel in concrete. Res. Chem. Intermediate.
42 (2016) 7823-7840. https://doi.org/10.1007/s11164-016-2564-1

[111] I.S. Pradipta, D.L.Y. Kong, J.T.B. Lee, Corrosion inhibition of green tea extract on
steel reinforcing bar embedded in mortar, 14th International Conference on Concrete
Engineering and Technology, IOP Conf. Series: Mater. Sci. Eng. 431 (2018) 042013.
https://doi.org/10.1088/1757-899X/431/4/042013

[112] J.H. Sinduja, G.G. Kumar, Effect of green corrosion inhibitors on the corrosion
behaviour of reinforced concrete, Int. Res. J. Eng. Tech. 6 (2019) 1656-1661.

[113] F. Wu, J. Xia, X. Liu, W. Zhai, X. Yuan, J. Yao, X. Xiao, Corrosion inhibition of
gatifloxacin on mild steel in 3% HCl solution, Corros. Sci. Prot. Technol. 28 (2016)
560-564.

[114] E.E Oguzie, Evaluation of the inhibitive effect of some plant extracts on the acid
corrosion of mild steel, Corros. Sci. 50 (2008) 2993-2998.
https://doi.org/10.1016/j.corsci.2008.08.004

[115] X.D. Zhang, G. Yv, F. Xv, Research on inhibition effect of multi-function pickling
corrosion inhibitor, Contemp. Chem. Ind. 2 (2016) 279-281.

[116] E.R. Clemente, J.G. Gonzalez-Rodriguez, G Valladarez, G.F. Dominguez-Patino,
Corrosion inhibition of carbon steel in H_2SO_4 by Chenopodiumambrosioides, Int. J.
Electrochem. Sci. 6 (2011) 6360-6372.

[117] X. Wang, F.Y. Wang, Y.X. Chen, X.X. Ji, S.C. Shi, Research progress of plant extract as green corrosion inhibitor, Corros. Sci. Prot. Technol. 29 (2017) 85-90.

[118] J. Zheng, Y.N. Wang, B.L. Zhang, T. Yang, X.M. Jiao, Y. He, Research progress on natural products as corrosion inhibitor in acid, neutral and alkaline mediums, Corros. Sci. Prot. Technol. 23 (2011) 103-106.

[119] E. Salehi, R. Naderi, B. Ramezanzadeh, Synthesis and characterization of an effective organic/inorganic hybrid green corrosion inhibitive complex based on zinc acetate/Urtica Dioica, Appl. Surf. Sci. 396 (2017)1499-1514. https://doi.org/10.1016/j.apsusc.2016.11.198

[120] L.R. Chauhan, G. Gunasekaran, Corrosion inhibition of mild steel by plant extract in dilute HCl medium, Corros. Sci. 49 (2007) 1143-1161. https://doi.org/10.1016/j.corsci.2006.08.012

[121] P.C. Okafor, M.E. Ikpi, I.E. Uwah, E.E. Ebenso, U.J. Ekpe, S.A. Umoren, Inhibitory action of Phyllanthus amarus extracts on the corrosion of mild steel in acidic media, Corros. Sci. 50 (2008) 2310-2317. https://doi.org/10.1016/j.corsci.2008.05.009

[122] A. Krishnan, S.M.A. Shibli, Optimization of an efficient, economic and eco-friendly inhibitor based on Sesbania grandiflora leaf extract for the mild steel corrosion in aggressive HCl environment, Anti-Corros. Methods Mater. 65 (2018) 210-216. https://doi.org/10.1108/ACMM-06-2017-1810

[123] H. Gerengi, H.I. Sahin, Schinopsis lorentzii extract as a green corrosion inhibitor for low carbon steel in 1 M HCl solution, Ind. Eng. Chem. Res. 51 (2012) 780-787. https://doi.org/10.1021/ie201776q

[124] M. Soudani, M.H. Meliani, K. El-Miloudi, O. Bouledroua, C. Fares, M.A. Benghalia, Z. Azari, J. Capelle, Ahmad A. Sorour, G. Pluvinage, Efficiency of green inhibitors against hydrogen embrittlement on mechanical properties of pipe steel API 5L X52 in hydrochloric acid medium, J. Bio. Tribo. Corros. 4 (2018) 36. https://doi.org/10.1007/s40735-018-0153-0

[125] X.H. Li, S.R. Deng, X.G. Xie, G.B. Du, Inhibition effect of bamboo leaf extract on the corrosion of aluminum in HCl solution, Acta Phys. Chim. Sin. 30(2014)1883-1894.

[126] W. Syu, M. Don, G. Lee, C. Sun, Cytotoxic and novel compounds from solanum indicum, J. Nat. Prod. 64 (2001) 1232-1233. https://doi.org/10.1021/np010186v

[127] M. El-Aasr, H. Miyashita, T. Ikeda, J. Lee, H. Yoshimitsu , T. Nohara, K. Murakami, A new spirostanol glycoside from fruits of solanum indicum L, Chem. Pharm. Bull. 57 (2009) 747-748. https://doi.org/10.1248/cpb.57.747

[128] X. Wang, H. Jiang, D. Zhang, L. Hou, W. Zhou, Solanum lasiocarpum *L*. Extract as green corrosion inhibitor for A3 steel in 1 M HCl solution, Int. J. Electrochem. Sci. 14 (2019)1178-1196. https://doi.org/10.20964/2019.02.06

[129] R. Idouhli, Y. Koumya, M. Khadiri, A. Aityoub, A. Abouelfida, A. Benyaich, Inhibitory effect of Senecio anteuphorbiumas green corrosion inhibitor for S300 steel, Int. J. Ind. Chem. 10 (2019) 133-143. https://doi.org/10.1007/s40090-019-0179-2

[130] N. Hassan, S.M. Ali, A. Ebrahim, H. El-Adwy, Performance evaluation and optimization of Camellia sinensis extract as green corrosion inhibitor for mild steel in acidic medium, Mater. Res. Express 6 (2019) 0865c7. https://doi.org/10.1088/2053-1591/ab2376

[131] J.R. Davis, Surface Engineering for Corrosion and Wear Resistance, ASM International, Cleveland, OH, USA, 2001.

[132] Y. Li, L. Li, J. Yu, Applications of zeolites in sustainable chemistry, Chem. 3 (2017) 928-949. https://doi.org/10.1016/j.chempr.2017.10.009

[133] M.F. Montemor, Functional and smart coatings for corrosion protection: A review of recent advances, Surf. Coat. Technol. 258 (2014) 17-37. https://doi.org/10.1016/j.surfcoat.2014.06.031

[134] H. Zheng, Y. Shao, Y. Wang, G. Meng, B. Liu, Reinforcing the corrosion protection property of epoxy coating by using graphene oxide-poly (urea-formaldehyde) composites, Corros. Sci. 123 (2017) 267-277. https://doi.org/10.1016/j.corsci.2017.04.019

[135] N.P. Tavandashti, M. Ghorbani, A. Shojaei, J.M.C. Mol, H. Terryn, K. Baert, Y. Gonzalez-Garcia, Inhibitor-loaded conducting polymer capsules for active corrosion protection of coating defects, Corros. Sci. 112 (2016), 138-149. https://doi.org/10.1016/j.corsci.2016.07.003

[136] V.S. Sastri, Green Corrosion Inhibitors: Theory and Practice, Wiley, Singapore, 2011. https://doi.org/10.1002/9781118015438

[137] M. Forsyth, B. Hinton, Rare Earth-Based Corrosion Inhibitors, Woodhead Publishing, Cambridge, OK, USA, 2014.

[138] K.A. Yasakau, M.L. Zheludkevich, S.V. Lamaka, M.G. Ferreira, Mechanism of corrosion inhibition of AA2024 by rare-earth compounds, J. Phys. Chem. B 110 (2006) 5515-5528. https://doi.org/10.1021/jp0560664

[139] M. Garcia-Heras, A. Jimenez-Morales, B. Casal, J.C. Galvan, S. Radzki, M.A. Villegas, Preparation and electrochemical study of cerium-silica sol-gel thin films. J. Alloys Compd. 380 (2004) 219-224. https://doi.org/10.1016/j.jallcom.2004.03.047

[140] A.N. Khramov, N.N. Voevodin, V.N. Balbyshev, R.A. Mantz, Sol-gel-derived corrosion-protective coatings with controllable release of incorporated organic corrosion inhibitors, Thin Solid Films 483 (2005)191-196. https://doi.org/10.1016/j.tsf.2004.12.021

[141] D.G. Shchukin, M. Zheludkevich, K. Yasakau, S. Lamaka, M.G.S. Ferreira, H. Mohwald, Layer-by-layer assembled nanocontainers for self-healing corrosion protection, Adv. Mater. 18 (2006)1672-1678. https://doi.org/10.1002/adma.200502053

[142] D. Grigoriev, E.Shchukina, D.G. Shchukin, Nanocontainers for self-healing coatings, Adv. Mater. Interfaces 4 (2017) 1-11. https://doi.org/10.1002/admi.201600318

[143] W. Yan, W.K. Ong, L.Y. Wu, S.L. Wijesinghe, Investigation of using sol-gel technology for corrosion protection coating systems incorporating colours and inhibitors, Coatings 9 (2019)1-15. https://doi.org/10.3390/coatings9010052

[144] T. Wang, J. Du, S. Ye, L. Tan, J. Fu, Triple-stimuli-responsive smart nanocontainers enhanced self-healing anticorrosion coatings for protection of aluminum alloy, ACS Appl. Mater. Interfaces 11 (2019) 4425-4438. https://doi.org/10.1021/acsami.8b19950

[145] N.J. Kanu, E. Gupta, U.K. Vates, G.K. Singh, Self-healing composites: A state-of-the-art review, Compos. A Appl. Sci. Manuf. 121 (2019) 474-486. https://doi.org/10.1016/j.compositesa.2019.04.012

[146] M.D. Brabander, H.R. Fischer, S.J. Garcia, Self-healing polymeric systems: Concepts and applications, in: M.R. Aguilar, J.S. Roman (Eds.), Smart Polymers and their Applications, Woodhead Publishing, Sawston, UK, 2019, pp. 379-409. https://doi.org/10.1016/B978-0-08-102416-4.00011-9

[147] R. Samiee, B. Ramezanzadeh, M. Mahdavian, E. Alibakhshi, Assessment of the smart self-healing corrosion protection properties of a water-base hybrid organo-silane film combined with non-toxic organic/inorganic environmentally friendly

corrosion inhibitors on mild steel, J. Clean. Prod. 220 (2019) 340-356.
https://doi.org/10.1016/j.jclepro.2019.02.149

[148] X. Zhao, Y. Li, B. Li, T. Hu, Y. Yang, L. Li, J. Zhang, Environmentally benign
and durable superhydrophobic coatings based on SiO₂ nanoparticles and silanes, J.
Colloid. Interface Sci. 542 (2019) 8-14. https://doi.org/10.1016/j.jcis.2019.01.115

[149] Y. Liu, H. Gu, Y. Jia, J. Liu, H. Zhang, R. Wang, B. Zhang, H. Zhang, Q. Zhang,
Design and preparation of biomimetic polydimethylsiloxane (PDMS) films with
superhydrophobic, self-healing and drag reduction properties via replication of shark
skin and SI-ATRP, Chem. Eng. J. 356 (2019) 318-328.
https://doi.org/10.1016/j.cej.2018.09.022

[150] S.L. de Armentia, M. Pantoja, J.Abenojar, M. Martinez, S.L. de Armentia, M.
Pantoja, J.Abenojar, M.A. Martinez, Development of silane-based coatings with
zirconia nanoparticles combining wetting, tribological, and aesthetical
properties, Coatings 8 (2018) 368. https://doi.org/10.3390/coatings8100368

[151] L. Ma, F. Svec, T. Tan, Y. Lv, In-situ growth of highly permeable zeolite
imidazolate framework membranes on porous polymer substrate using metal chelated
polyaniline as interface layer, J. Membrane Sci. 576 (2019) 1-8.
https://doi.org/10.1016/j.memsci.2019.01.011

[152] Y. Tan, F. Wang, J. Zhang, Design and synthesis of multifunctional metal-organic
zeolites, Chem. Soc. Rev. 47 (2018) 2130-2144.
https://doi.org/10.1039/C7CS00782E

[153] J. Wang, X. Song, J. Wang, X. Cui, Q. Zhou, T. Qi, G.L. Li, Smart-sensing
polymer coatings with autonomously reporting corrosion dynamics of self-healing
systems, Adv. Mater. Interfaces 6 (2019) 1900055.
https://doi.org/10.1002/admi.201900055

[154] Z. Karami, S. Maleki, A. Moghaddam, A. Jahandideh, Self-healing bio-
composites: Concepts, developments and perspective, in: Inamuddin, S. Thomas,
R.K. Mishra, A.M. Asiri (Eds.), Sustainable Polymer Composites and
Nanocomposites, Springer, Cham, Switzerland, 2019, pp. 1323-1343.
https://doi.org/10.1007/978-3-030-05399-4_44

[155] M. Catauro, S.V. Ciprioti, Sol-gel synthesis and characterization of hybrid
materials for biomedical applications, in: C. Demetzos, N. Pippa (Eds.),
Thermodynamics and Biophysics of Biomedical Nanosystems, Springer: Singapore,
2019, pp. 445–475. https://doi.org/10.1007/978-981-13-0989-2_13

[156] D. Prasai, J.C. Tuberquia, R.R. Harl, G.K. Jennings, K.I. Bolotin, Graphene: Corrosion-inhibiting coating, ACS Nano 6 (2012) 1102–1108. https://doi.org/10.1021/nn203507y

[157] Y. Zhu, S. Murali, W. Cai, X. Li, J.W. Suk, J.R. Potts, R.S. Ruoff, Graphene and graphene oxide: Synthesis, properties, and applications, Adv. Mater. 22 (2010) 3906-3924. https://doi.org/10.1002/adma.201001068

[158] C. Lee, X. Wei, J.W. Kysar, J. Hone, Measurement of the elastic properties and intrinsic strength of monolayer graphene, Science 321 (2008) 385-388. https://doi.org/10.1126/science.1157996

[159] A.A. Balandin, S. Ghosh, W. Bao, I. Calizo, D. Teweldebrhan, F. Miao, C.N. Lau, Superior thermal conductivity of single-layer graphene, Nano Lett. 8 (2008) 902-907. https://doi.org/10.1021/nl0731872

[160] H. Song, L. Zhang, C. He, Y. Qu, Y. Tian, Y. Lv, Graphene sheets decorated with SnO_2 nanoparticles: In situ synthesis and highly efficient materials for cataluminescence gas sensors, J. Mater. Chem. 21 (2011) 5972-5977. https://doi.org/10.1039/c0jm04331a

[161] M. Liang, L. Zhi, Graphene-based electrode materials for rechargeable lithium batteries, J. Mater. Chem. 19 (2009) 5871-5878. https://doi.org/10.1039/b901551e

[162] A.L.M. Reddy, A. Srivastava, S.R. Gowda, H. Gullapalli, M. Dubey, P.M. Ajayan, Synthesis of nitrogen-doped graphene films for lithium battery application, ACS Nano 4 (2010) 6337-6342. https://doi.org/10.1021/nn101926g

[163] J. Cheng, G. Zhang, J. Du, L. Tang, J. Xu, J. Li, New role of graphene oxide as active hydrogen donor in the recyclable palladium nanoparticles catalyzed ullmann reaction in environmental friendly ionic liquid/supercritical carbon dioxide system, J. Mater. Chem. 21 (2011) 3485-3494. https://doi.org/10.1039/c0jm02396e

[164] G. Kalita, M. Matsushima, H. Uchida, K. Wakita, M. Umeno, Graphene constructed carbon thin films as transparent electrodes for solar cell applications, J. Mater. Chem. 20 (2010) 9713-9717. https://doi.org/10.1039/c0jm01352h

[165] K. Zhang, L.L. Zhang, X.S. Zhao, J. Wu, Graphene/polyaniline nanofiber composites as supercapacitor electrodes, Chem. Mater. 22 (2010) 1392-1401. https://doi.org/10.1021/cm902876u

[166] T. Ramanathan, A.A. Abdala, S. Stankovich, D.A. Dikin, M.H. Alonso, R.D. Piner, D.H. Adamson, H.C. Schniepp, X. Chen, R.S. Ruoff, Functionalized graphene

sheets for polymer nanocomposites, Nat. Nanotechnol. 3 (2008) 327-331.
https://doi.org/10.1038/nnano.2008.96

[167] D. Cai, M. Song, Recent advance in functionalized graphene/polymer
nanocomposites, J. Mater. Chem. 20 (2010) 7906-7915.
https://doi.org/10.1039/c0jm00530d

[168] H.J. Salavagione, G. Martinez, M.A. Gomez, Synthesis of poly(vinyl
alcohol)/reduced graphite oxide nanocomposites with improved thermal and
electrical properties, J. Mater. Chem. 19 (2009) 5027-5032.
https://doi.org/10.1039/b904232f

[169] J. Zhao, S. Pei, W. Ren, L. Gao, H.M. Cheng, Efficient preparation of large-area
graphene oxide sheets for transparent conductive films, ACS Nano 4 (2010) 5245-
5252. https://doi.org/10.1021/nn1015506

[170] J. Zhu, Graphene production: New solutions to a new problem, Nat Nano 3 (2008)
528-529. https://doi.org/10.1038/nnano.2008.249

[171] C.H. Chang, T.C. Huang, C.W. Peng, T.C. Yeh, H.I. Lu, W.I. Hung, C.J. Weng,
T.I. Yeh, J.M. Yang, Novel anticorrosion coatings prepared from
polyaniline/graphene composites, Carbon 50 (2012) 5044-5051.
https://doi.org/10.1016/j.carbon.2012.06.043

[172] J. Qiu, S. Wang, Enhancing polymer performance through graphene sheets, J.
Appl. Polym. Sci. 119 (2011) 3670-3674. https://doi.org/10.1002/app.33068

[173] A. Yasmin, I.M. Daniel, Mechanical and thermal properties of graphite
platelet/epoxy composites, Polymer 45 (2004) 8211-8219.
https://doi.org/10.1016/j.polymer.2004.09.054

[174] M.A. Rafiee, J. Rafiee, Z. Wang, H. Song, Z.Z. Yu, N. Koratkar, Enhanced
mechanical properties of nanocomposites at low graphene content, ACS Nano 3
(2009) 3884-3890. https://doi.org/10.1021/nn9010472

[175] S.G. Miller, J.L. Bauer, M.J. Maryanski, P.J. Heimann, J.P. Barlow, J.M. Gosau,
R.E. Allred, Characterization of epoxy functionalized graphite nanoparticles and the
physical properties of epoxy matrix nanocomposites, Compos. Sci. Technol. 70
(2010) 1120-1125. https://doi.org/10.1016/j.compscitech.2010.02.023

[176] C.L. Chiang, S.W. Hsu, Synthesis, characterization and thermal properties of
novel epoxy/expandable graphite composites, Polym. Int. 59 (2010) 119-126.
https://doi.org/10.1002/pi.2699

[177] M. Martin-Gallego, R. Verdejo, M.A. Lopez-Manchado, M. Sangermano, Epoxy-graphene UV-cured nanocomposites, Polymer 52 (2011) 4664-4669. https://doi.org/10.1016/j.polymer.2011.08.039

[178] P.A. Sorensen, S. Kiil, K. Dam-Johansen, C.E. Weinell, Anticorrosive coatings: A review, J. Coat. Technol. Res. 6 (2009) 135-176. https://doi.org/10.1007/s11998-008-9144-2

[179] C. Bao, Y. Guo, L. Song, Y. Kan, X. Qian, Y. Hu, In situ preparation of functionalized graphene oxide/epoxy nanocomposites with effective reinforcements, J. Mater. Chem. 21 (2011) 13290-13298. https://doi.org/10.1039/c1jm11434d

[180] H. Fang, K. Chen, Z. Zhang, C. Zhu, Effect of Yb addition on microstructure and properties of 7A60 aluminium alloy, Trans. Nonferrous Met. Soc. China. 18 (2008) 28-32. https://doi.org/10.1016/S1003-6326(08)60006-0

[181] S. Caporali, A. Fossati, A. Lavacchi, I. Perissi, A. Tolstogouzov, U. Bardi, Aluminium electroplated from ionic liquids as protective coating against steel corrosion, Corros. Sci., 50 (2008) 534-539. https://doi.org/10.1016/j.corsci.2007.08.001

[182] C. Verma, P. Singh, I. Bahadur, E. Ebenso, M.A. Quraishi, Electrochemical, thermodynamic, surface and theoretical investigation of 2-aminobenzene-1,3-dicarbonitriles as green corrosion inhibitor for aluminum in 0.5 M NaOH, J. Mol. Liq. 209 (2015) 767-778. https://doi.org/10.1016/j.molliq.2015.06.039

[183] T.C.E. Tringham, Causes and Prevention of Corrosion in Aircraft, Pitman, London, 1958.

[184] S. Benavides, Corrosion Control in the Aerospace Industry, first ed., CRC Press, USA, 2009. https://doi.org/10.1201/9781439829202.ch1

[185] K. Xhanari, M. Finsgar, M.K. Hrncic, U. Maver, Z. Kneza, B. Seiti, Green corrosion inhibitors for aluminium and its alloys: A review, RSC Adv. 7 (2017) 27299-27330. https://doi.org/10.1039/C7RA03944A

[186] N. Raghavendra, J.I. Bhat, Anti-corrosion Properties of Areca Palm Leaf Extract on Aluminium in 0.5 M HCl Environment, S. Afr. J. Chem. 71 (2018) 30-38. https://doi.org/10.17159/0379-4350/2018/v71a4

[187] O.U. Abakedi, Inhibition of aluminium corrosion in hydrochloric acid solution by Stachytarpheta indica leaf extract, J. Sci. Eng. Res. 3 (2016) 105-110.

[188] O.U. Abakedi, I.E. Moses, Aluminium corrosion inhibition by maesobatrya barteri root extract in hydrochloric acid solution, Am. Chem. Sci. J. 10 (2016) 1-10. https://doi.org/10.9734/ACSJ/2016/21812

[189] A. Singh, I. Ahamad, M.A. Quraishi, Piper longum extract as green corrosion inhibitor for aluminium in NaOH solution, Arab. J. Chem. 9 (2016) S1584-S1589. https://doi.org/10.1016/j.arabjc.2012.04.029

[190] L.A. Nnanna, K.O. Uchendu, G. Ikwuagwu, W.O. John, U. Ihekoronye, Inhibition of corrosion of aluminum alloy AA8011 in alkaline medium using palisota hirsute extract, Int. Lett. Chem. Phys. Astron. 67 (2016) 14-20. https://doi.org/10.18052/www.scipress.com/ILCPA.67.14

[191] N. Chaube, V.K. Singh, Savita, M.A. Quraishi, E.E. Ebenso, Corrosion inhibition of aluminium alloy in alkaline media by neolamarkia cadamba bark extract as a green inhibitor, Int. J. Electrochem. Sci.10 (2015) 504-518.

[192] S. Deng, X. Li, Inhibition by *jasminum nudiflorum* lindl. leaves extract of the corrosion of aluminium in HCl solution, Corros. Sci. 64 (2012) 253-262. https://doi.org/10.1016/j.corsci.2012.07.017

[193] K. Krishnaveni, J. Ravichandran, Effect of aqueous extract of leaves of Morinda tinctoria on corrosion inhibition of aluminium surface in HCl medium, T. Nonferr. Metal Soc. 24 (2014) 2704-2712. https://doi.org/10.1016/S1003-6326(14)63401-4

[194] X. Li, S. Deng, Inhibition effect of Dendrocalamus brandisii leaves extract on aluminumin HCl, H_3PO_4 solutions, Corros. Sci. 65 (2012) 299-308. https://doi.org/10.1016/j.corsci.2012.08.033

[195] A. Singh, E.E. Ebenso, M.A. Quraishi, Stem extract of brahmi (Bacopa Monnieri) as green corrosion inhibitor for aluminum in NaOH solution, Int. J. Electrochem. Sci, 7 (2012) 3409-3419.

[196] A. Ali, N. Foaud, Inhibition of aluminum corrosion in hydrochloric acid solution using black mulberry extract, J. Mater. Environ. Sci, 3 (2012) 917-924.

[197] J. Halambek, K. Berkovic, J. Vorkapic-Furac, *Laurusnobilis* L. oil as green corrosion inhibitor for aluminium and AA5754 aluminium alloy in 3% NaCl solution, Mater. Chem. Phys. 137 (2013) 788-795. https://doi.org/10.1016/j.matchemphys.2012.09.066

[198] I.M. Mejeha, M.C. Nwandu, K.B. Okeoma, L.A. Nnanna, M.A. Chidiebere, F.C. Eze, E.E. Oguzie, Experimental and theoretical assessment of the inhibiting action of

Aspilia Africana extract on corrosion aluminium alloy AA3003 in hydrochloric acid, J. Mater. Sci. 47 (2012) 2559-2572. https://doi.org/10.1007/s10853-011-6079-2

[199] S. Ambrish, M.A. Quraishi, Azwain (Trachyspermum copticum) seed extract as an efficient corrosion inhibitor for aluminium in NaOH solution, Res. J. Recent Sci. 1 (2012) 57-61.

[200] L.A. Nnanna, V. Obasi, O. Nwadiuko, K. Mejeh, N. Ekekwe, S.C. Udensi, Inhibition by Newbouldia leavis leaf extract of the corrosion of aluminium in HCl and H_2SO_4 solutions, Arch. Appl. Sci. Res. 4 (2012) 207-217.

[201] L.A. Nnanna, O.C. Nwadiuko, N.D. Ekekwe, C.F. Ukpabi, S.C. Udensi, K.B. Okeoma, B.N. Onwuagba, I.M. Mejeha, Adsorption and inhibitive properties of leaf extract of Newbouldia leavis as a green inhibitor for aluminium alloy in H_2SO_4, Am. J. Mater. Sci. 1 (2011) 143-148. https://doi.org/10.5923/j.materials.20110102.24

[202] H.M. Elabbasy, A.S. Fouda, Olive leaf as green corrosion inhibitor for C-steel in Sulfamic acid solution, Green Chem. Lett. Rev. 12 (2019) 332-342. https://doi.org/10.1080/17518253.2019.1646812

[203] M. Deyab, Inhibition activity of Seaweed extract for mild carbon steel corrosion in saline formation water, Desalination 384 (2016) 60-67. https://doi.org/10.1016/j.desal.2016.02.001

[204] A. El Bribri, M. Tabyaoui, B. Tabyaoui, H. El Attari, F. Bentiss, The use of Euphorbia falcata extract as eco-friendly corrosion inhibitor of carbon steel in hydrochloric acid solution, Mater. Chem. Phys. 141 (2013) 240-247. https://doi.org/10.1016/j.matchemphys.2013.05.006

[205] M. Deyab, Egyptian licorice extract as a green corrosion inhibitor for copper in hydrochloric acid solution, J. Ind. Eng. Chem. 22 (2015) 384-389. https://doi.org/10.1016/j.jiec.2014.07.036

Theory and Applications of Green Corrosion Inhibitors
Materials Research Forum LLC
Materials Research Foundations **86** (2021) 161-182
https://doi.org/10.21741/9781644901052-6

Chapter 6

Pyrazine Derivatives as Green Corrosion Inhibitors

A.K. Dewangan, Y. Dewangan, D.K. Verma*

Department of Chemistry, Government Digvijay Autonomous Postgraduate College, Rajnandgaon, Chhattisgarh, India, 491441

*dakeshwarverma@gmail.com

Abstract

Pyrazine is a six-member heterocyclic compound containing two nitrogen atoms, located at one-fourth position. A pyrazine and its derivatives are most commonly used as a pharmaceutical, food colouring agent and as a corrosion inhibitor. It has chemistry and chemical characteristics due to nitrogen atoms present which can be obtained from both naturally and synthetically. A pyrazine and its derivatives are used effectively in corrosion media to prevent metal and metal alloy from corrosion. Chemical and electrochemical methods are commonly used as corrosion monitoring techniques, in which weight loss EIS and PDP are prominent. Their mixed type of nature has been reported in previously published works. There have also been works on the use of computation calculation in recent works, which include DFT calculation and MD Simulation are the main ones. The future research related to the pyrazine has also been highlighted.

Keywords

Pyrazine, Metals, Solution, Computational, Electrochemical

Contents

1.   Introduction

Currently, metal and metal alloys are used in almost all types of industries such as pharmaceutical, oil –gas, aeronautics, construction and petroleum [1]. The metal and alloys immersed in an aggressive solution (acidic/basic) during the acid pickling and scaling process so that the deposit rust can be removed over them. There is a loss of rust as well as base metal during the rust cleaning [2]. This loss of metal is caused by the aggressive ion attacking the metal surface. A small amount of inhibitors is used in corrosive solutions to prevent this loss, by adsorbing on the metal surface [3]. The inhibitors used can be both natural and synthetic. Generality: hetero atoms and π bond containing organic compounds are placed in the category of good inhibitors because lone pair and/or π - electrons present in them can easily share their electron with metal [4-6]. In this order, the use of nitrogen-containing heterogeneous cyclic compounds such as triazole derivative [7]. indole derivative [8,9], pyridine derivative [10-12], tetrazole derivative [13] quinoxaline [14,15] pthalazine [16], pyradazine derivatives [17,18], pyrazole derivatives [19] has been done as corrosion inhibitors. Research has shown the good corrosion inhibitor property of Nitrogen-containing heterocyclic organic molecules. The study of di nitrogen-containing pyrazine derivatives has also been prominent. Pyrazine derivatives with nitrogen atom, pi bonds as well as substituted group show prominent corrosion inhibitor towards metal corrosion. A pyrazine and its derivative are one of the important heterocyclic compounds that can be synthesized by chemical reaction or it can be obtained from natural sources [20,21]. These are electron deficient due to the inductive effect, this adjacent developed partially positive charge on the carbon atom, because of which the nucleophile in them attacks the carbon atoms.

In this way, various derivatives are synthesized. Pyrazine derivative is used prominently in the food industry and drug materials such as antimicrobial antiviral, antiulcer etc. [22-24]. Overall, this book chapter describes the use of pyrazine-based potential corrosion inhibitor. The unique property of pyrazine such as solubility, less toxic, cost-effective and biodegradability makes it suitable as a corrosion inhibitor. Theoretical analysis such as

Theory and Applications of Green Corrosion Inhibitors   Materials Research Forum LLC
Materials Research Foundations **86** (2021) 161-182   https://doi.org/10.21741/9781644901052-6

DFT, MD simulations and DFT calculations and electrochemical methods like PDP, EIS and OCP have been mainly reported in published work to determine the inhibition property. The previous study suggests that EIS, PDP, OCP, applied for electrochemical analysis in corrosive media such as HCl, HNO_3, H_2SO_4, NaCl, etc. Other methods like weight loss, SEM-EDS employed to support the electrochemical study. Fig. 1 represents the schematic representation of methods extensively used for corrosion monitoring in metal and alloys at various aggressive media.

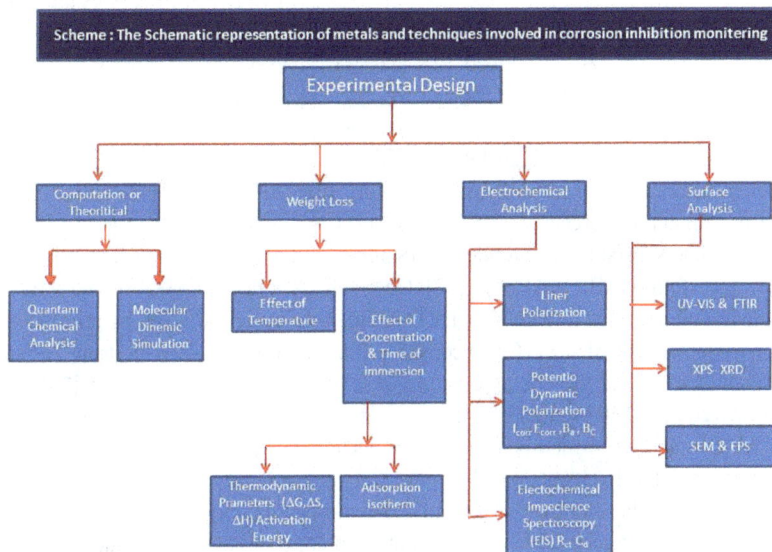

Figure 1. Schematic representation of corrosion monitoring methods.

2.   Pyrazine and its derivative as prominent corrosion inhibitor for metals and alloys in corrosive media

Limited research papers on pyrazine as corrosion inhibitor are published. Generally, EIS and PDP applied as corrosion inhibitor monitoring techniques for electrochemical analysis. SEM-EDS used for surface analysis and determining the composition and theoretical study such as MD, simulation and DFT calculation used organically to determine the chemical behaviour of the organic molecules. Pyrazine derivative namely Pyrazine C, Pyrazine E, Pyrazine H used as corrosion inhibitor for API 60 steel studied using various techniques like EDS, LPR, EIS, WL in which Pyrazine C exhibited higher % inhibition efficiency *(IE)* of 90.04 at 1% Concentration with NaI and Glu [25]. Hadi and co-workers reported three derivatives namely MP, AP and ABP applied for the protection of corrosion in which ABP considered as good inhibitor among the studied inhibitors [26]. DFT calculation reveals the good correlation with all studied parameters that inhibitors work efficiently towards metal corrosion [27]. PDP study reveals the % *IE* of 96.85% and 58.35% respectively for CPP and PP at 35°C temperature [28]. Daobing haung and co-workers (2013) applied EIS and WL analysis, which reveals the good efficiency of pyrazine towards corrosion of Mg, Zr, Gd composites [29]. J. Saranaya et al. reported that APP as an efficiency corrosion inhibitor and show good % *IE* of 75.82% (10 mM) according to PDP analysis [30]. Chemical and Electrochemical analysis reveals the efficient nature of DPD towards mild steel corrosion (82% IE) in 10-2 M at 25°C as per M. Bouklah and researchers (2005) analysis [31]. According to Xianghong Li (2013) and co-researchers in studied inhibitors, EIS reveals that the order towards corrosion inhibition is as follow ABP > AP>MP [32]. Obot et al. reported the MP, AP and ABP as corrosion inhibitor for steel in which DFT calculation suggested ABP as an efficient inhibitor [33]. Cold rolled steel (CRS) corrosion inhibition was studied by using AP and ABP. Techniques reveal the effective nature of inhibitions towards corrosion [34]. EIS analysis proved that DPB worked efficiency as corrosion inhibition and shows 80 % *IE* in 5 ×10^{-2} M ppm at 25°C which is supported by OCP analysis [35]. Pyrazine derivatives namely DP-I and DP-II characterised as potential corrosion inhibitor using EDS, DFT, EIS, WL and PDP where EIS showed that DP-I and DP-II exhibited inhibition efficiency of 98.30% and 83.35% respectively at their optimum concentration [36]. Shuo Zang et al (2016) applied EDS, EIS and SEM techniques where studied parameters proved the good inhibitory property (89.9%) of PD-I towards the mild steel corrosion at 1000mg/ concentration [37]. Studied parameters EIS and OCP proved the good inhibitory property of inhibitor towards the mild steel corrosion at 1.5 mol/L concentration. Synergistic effect between PCA and KI enhance the inhibition efficiency up to 92.6 % at an equal concentration as per PDP analysis. It was reported the MP, AP and ABP as corrosion inhibitor for mild steel ABP proved as an efficient inhibitor according to MD simulation and DFT calculation [38]. EIS analysis showed that P1 and P2 exhibited 58.39% and 96.85 % inhibition efficiency at 1.25 mM concentration at 303 K temperature [39]. Silver containing pyrazine complexes synthesised and applied as prominent corrosion inhibition and found mixed type behaviour of both complex1 and complex 2 followed by Anodic effect predominantly. PDP analysis reveals 68% and 50% inhibition efficiency for the complex 1 and complex 2 respectively [40]. Pyrido [2,3,b]pyrazine derivative synthesis and used as a prominent corrosion inhibitor in aggressive solution (1M HCl) for steel corrosion and reveal the 93% inhibition efficiency at 10^{-3} M concentration of inhibition namely P1 [41-43]. 3 Pyrazine derivatives namely PDA, ACA and PA exhibit resistivity towards alloy corrosion in 3M HCl in which ACA shows maximum % *IE* of 85.5 % at e 11×10^{-6} M conc. As per EIS analysis [44], PDP analysis reveals that PQDPP exhibited the highest corrosion efficiency of 90.60% at 60ppm among Quinoxalin-6-yl pyrazine derivative (PQDPB, PQDPP, PPQDPE) [45]. Three Pyrado[2,3-b] pyrazine derivative (P1) proved as efficiency corrosion inhibitor showing 93% *IE* at 10^{-3} concentration as per EIS analysis [46]. Three Quinoxaline derivative namely STQ, BMQ and FVQ studied by various techniques like EIS, DFT, PDP as corrosion inhibitor for the mild steel / 1M HCl system in which their inhibition efficiency follow the order of BMQ (95.07 %) > FQV (93.74%) > STP (90.92%) [47]. Two Pyrazine derivative namely GP2 and GP3 applied for corrosion inhibitor for C38 steel (1HCl), in which GP3 shows 91.83 % inhibition efficiency and GP2 shows 83.90% inhibition efficiency according to gravimetric analysis at their optimum concentration of 1×10^{-3}M concentration [48]. Table.1 illustrates the molecular structure, metals and media, techniques applied, nature of inhibitors and outcomes of the work.

Table 1. The molecular structure, metals and media, techniques applied, nature of inhibitors and outcomes of the research

S. N.	Molecular structure	Metals & media	methods	Nature of inhibitor	Outcomes	Ref.
1	Pyrazine c Pyrazine e Pyrazine H	API X60 steel / 15% HCl	PDP, EFM, EIS, LPR, SEM, FTIR, EDX	Mixed / Langmuir model	Pyrazine derivative namely Pyrazine C, Pyrazine E, Pyrazine H used corrosion inhibitor API 60 steel studied using various-various techniques in which Pyrazine C exhibited higher %*IE* of 90.04 at 1% Concentration with NaI and Glu	25

2	MP AP ABP	MS / 1.0 M H$_2$SO$_4$	NICS, FLU, PDI, DFT, QUSR	-	Three derivative MP, AP and ABP applied for the protection of corrosion in which ABP considered as good inhibitor among the studied inhibitor	26
3	MP AP ABP	MS / 1.0 M H$_2$SO$_4$	DFT		DFT calculation reveals the good correlation with all studied parameters that inhibitors work efficiently towards metal corrosion	27
4	PDP CPP	MS / 0.5 M HCl	EIS, FTIR, UV-VIS	Mixed / Langmuir model	PDP study reveals the % *IE* of 96.85% and 58.35% respectively for CPP and PP at 35 °C temperature.	28
5		Mg-Zr-Gd alloy / Etane-1,2-diol	EIS, WL	-	EIS & WL analysis reveals the good efficiency of pyrazine towards corrosion of Mg, Zr, Gd composites	29
6	APP APQ	MS / 1.0 M H$_2$SO$_4$	EIS, SEM, AFM, DFT	Mixed / Langmuir model	J. Saranaya et. al. proved that APP acts as efficiency corrosion inhibitor and show good % *IE* of 75.82% (10 mM) according to PPQ analysis	30
7	DPD	MS / 0.5 M H$_2$SO$_4$	EIS, WL, PDP	Mixed / Langmuir model	Chemical and Electrochemical analysis reveals the efficient nature of DPD towards mild steel corrosion (82% IE) in 10^{-2} M at 25 ^{0}C	31

8	MP AP ABP	MS / 1.0 M H₂SO₄	EIS, WL, PDP	Mixed / Langmuir model	EIS reveals that ABP exhibited good inhibition property among studied inhibitor	32
9	MP AP ABP	Steel	DFT, MD simulati on	-	Obot et. al. applied MD simulation and DFT calculation in which DFT calculation suggested ABP as efficient inhibitor	33
10	AP ABP	CRS/ 1.0 M HCl	EIS, WL, PDP	Langmuir model	CRS corrosion inhibition study done by using AP and ABP. Techniques reveals the effective nature of inhibitions towards corrosion.	34
11	DPD	MS / 0.5 M H₂SO₄	EIS, OCP	Langmuir model	EIS analysis proved that DPB worked efficiency as corrosion inhibition and shows 80% *IE* in 5 × 10⁻² M ppm at 25 °C	35
12	DP I DP II	MS / 0.5 M H₂SO₄	EDS, DFT, PDP, EIS, WL	Langmuir model	DP-I and DP-II exhibited inhibition efficiency of 98.30% and 83.35% respectively at their optimum concentration.	36
13	PD I	MS / 15% HCl	EIS, SEM, EDX	Mixed / Langmuir model	Studied parameters proved the good inhibitory property (89.9%) of PD-I towards the mild steel corrosion at 1000mg/ concentration.	37

14	DPD	Steel / 0.5 M H_2SO_4	EIS, OCP	Langmuir model	EIS reveals the 80% *IE* at 5×10^{-3} M conc. of inhibitor, also supported the other techniques	38
15	PCA	CS / 0.5 M H_2SO_4	SEM, PDP	Mixed / Langmuir model	Synergistic effect between PCA and KI enhance the inhibition efficiency up to 92.6 % at equal concentration as per PDP analysis	39
16	MP AP ABP	MS / 1.0 M H_2SO_4	DFT, MD simulation		It was reported that MP, AP and ABP as corrosion inhibitor for mild steel ABP proved as efficient inhibitor according to MD simulation and DFT calculation	40
17	P I P II	MS / 0.5 M HCl	EIS, FTIR, UV-VIS	Langmuir model	EIS result shown that P1 and P2 exhibited 58.39% and 96.85 % inhibition efficiency at 1.25 mM concentration at 303 K temperature	41
18	AP PDA	MS / 0.1M HNO_3	X-ray Single crystal structure analysis (XSCSA), PDP, FTIR	Mixed / Langmuir model	Silver containing pyrazine complexes synthesised and applied as prominent corrosion inhibition and found mixed type behaviour of both complex1 and complex 2 followed by Anodic effect predominantly. PDP analysis reveals the 68% and 50% inhibition efficiency for the complex 1 and complex 2 respectively.	42

19	BAPPD	MS / 1M HCl	DFT, EIS	Langmuir model	Pyrido[2,3,b]pyrazine derivative (BAPPD) synthesis and used as prominent corrosion inhibitor in aggressive solution (1M HCl) for steel corrosion and reveal the 93% inhibition efficiency at 10^{-3} M concentration of inhibition namely P1	43
20	PDA APCA PA	Copper alloy (316 LSS) /3M HCl	WL, EIS, PDP	Temkin model	3 Pyrazine derivative namely PDA, ACA and PA exhibits resistivity towards alloy corrosion in 3M HCl in which ACA shows maximum % (*IE*) inhibition efficiency of 85.5 % at e 11X10^{-6} M conc. As per EIS analysis.	44
21	PQDBP PQDPP PPQDPE	MS / 1M HCl	OCP, EIS, PDP, SEM, DFT, MC Simulati on		L. Qlasonkanmi et al (2015) PDP analysis reveal that PQDPP exhibited highest corrosion efficiency of 90.60% at 60ppm among Quinoxalin-6-yl pyrazine derivatives (PQDPB, PQDPP, PPQDPE).	45

22	PI	MS / 1M HCl	WL, PDP, EIS, DFT	Langmuir Model	Pyrado[2,3-b] pyrazine derivative (P1) proved as efficiency corrosion inhibitor showing 93% inhibition efficiency at 10^{-3} concentration as per EIS analysis.	46
23	BMQ STQ FVQ	MS / 1M HCl	DFT, PDP, EIS	Langmuir model	Three quinoxaline derivative namely STQ, BMQ and FVQ studied by various techniques like EIS, DFT, PDP as corrosion inhibitor for the mild steel / 1M HCl system in which their inhibition efficiency follow the order of BMQ (95.07 %) > FQV (93.74%) > STP (90.92%)	47
24	GP 2 GP 3	C38 Steel/ 1M HCl	WL, PDP	Langmuir model	Two Pyrazine derivative namely GP2 and GP3 applied for corrosion inhibitor for C38 steel (1HCl), in which GP3 shows 91.83 % inhibition efficiency and GP2 shows 83.90% inhibition efficiency according to gravimetric analysis at their optimum concentration of 1×10^{-3}M concentration.	48

Theory and Applications of Green Corrosion Inhibitors   Materials Research Forum LLC
Materials Research Foundations **86** (2021) 161-182   https://doi.org/10.21741/9781644901052-6

3.   Adsorption mechanism

The adsorption process can be explained mainly by one or more of the given reasons:

(1) There is an interaction between the lone pair electron (s) and metal surface present in the atomic atom of the inhibitor molecule which can also be called chemistry of donor and accepter. Therefore, inhibitors share its electron to vacant d-orbital of a metal surface. This adsorption mechanism can be considered as chemical, physical, or both type of adsorption [49-51]. (2) Electrostatic interaction between metal and inhibitor molecules in which both metal and inhibitor are charged. In this mechanism, charged anions of acidic solutions such as Cl-, SO_4^{2-}, PO_4^{3-} , NO_3^- etc. deposit above the metal surface to form a permanent coordination intermediate. The H^+ obtained from the acidic solution interacts with the lone pair of heteroatoms and changes them into positive ion. Thus, through interaction between these charged molecules, the inhibitor molecules deposited onto metal [52-54]. Sometimes both these mechanisms are also reported where nature of inhibitors reported as cathodic, anodic, or mixed type [55-57]. The inhibitor normally deposited as a monolayer above the metal considered as Langmuir adsorption model [58-60]. Besides, inhibitor molecules follow Temkin, Freundlich adsorption have isotherm which has been reported in some published papers [61-63]. Generally, the nature of adsorption will be physical or chemical depending on the value of Gibbs free energy (ΔG) [64].   Figure 2 illustrated the approximate action method of the adsorption process between the pyrazine derivative and metal surface [25].

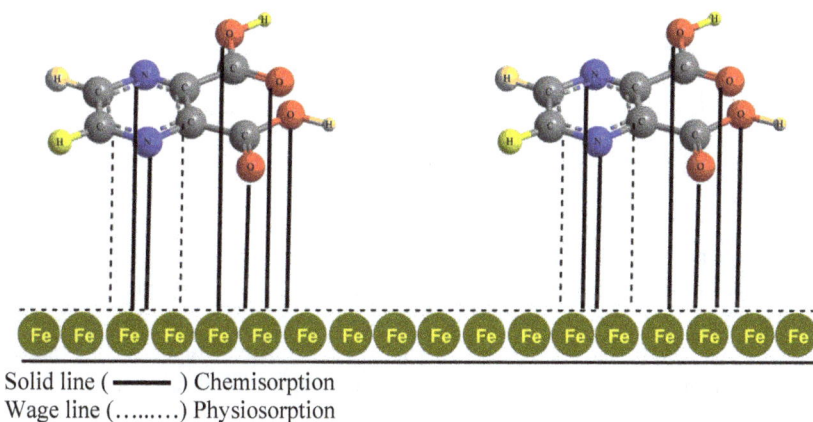

Solid line (────) Chemisorption
Wage line (.........) Physiosorption

Figure 2. An approximate action method of the adsorption process between pyrazine
derivative and metal surface is illustrated with the help of the diagram.

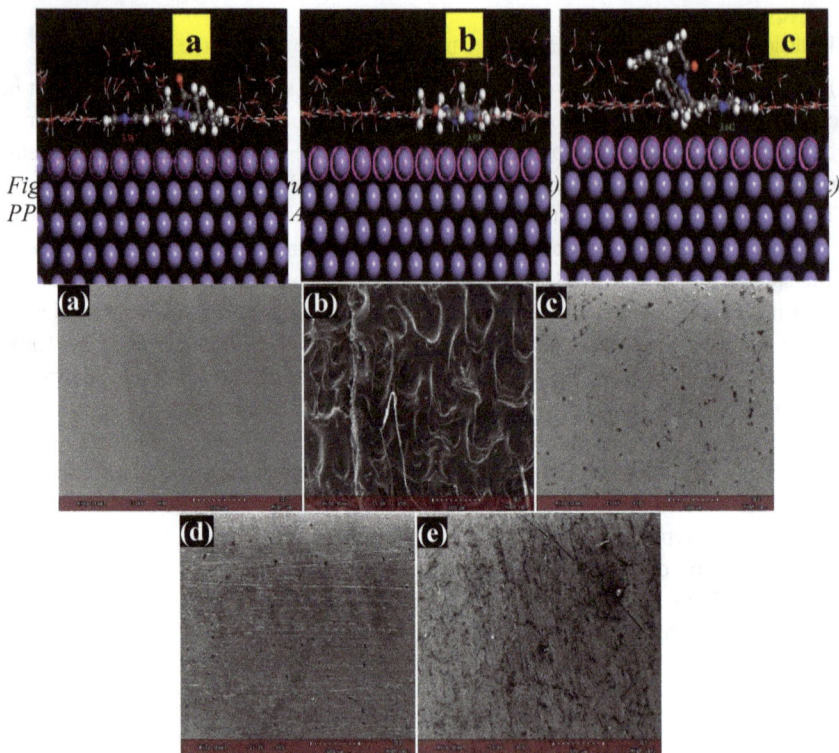

Figure 4. SEM images of mild steel (a) plain mild steel, (b) mild steel in 1 M HCl (blank), (c) PQDPB, (d) PQDPP and (e) PPQDPE [Copyright 2015@ American Chemical Society (ACS) Publication].

SEM analysis provide the information about inhibitor adsorption onto metal surface similarly theoretical analysis such MD simulation and DFT calculation also applied for determination of molecules properties as inhibitor adsorbed over metal surface. MD simulations or Molecular dynamics simulation is effective to analysis various macroscopic properties such as co-ordination number, radial distribution function, thickness, density, usual potential energy and dispersion of a system containing complex and poly molecules. Lukman Olasunkanmi and co-workers (2015) work on derivatives of three quinoxalin-6-yl derivatives namely PQDPB, PQDPP and PPQDPE as potential inhibitor on mild steel corrosion in aggressive acidic media (HCl). Among studied inhibitors PQDPP exhibited good corrosion resistant towards metal which is proved by techniques like SEM, quantum chemical analysis and MD simulation. MD simulation investigation reveals the horizontal as well as vertical (flat) alignment of studied molecules over metal by which they can adsorb on to mild steel surface as shown in Fig. 3. E_{int} values reflect the good relationship with other applied methods. Similarly, SEM microphotographs images shown in Fig. 4 reveals that in the pure acidic solution metal surface corroded badly while applying inhibitor less pitch and damage seen onto metal surface [47]. I.B. Obo and Z.M. Gasem applied theoretical calculation on some pyrazine derivatives for the evaluation of their molecular properties like LUMO, HOMO, total energy (ΔE) and dipole moment. These parameters are helpful to evaluate the inhibition property of organic molecules towards metal corrosion. Fig. 5 and 6 shows the HOMO, LUMO and optimised structure of some pyrazine derivative. Results revels the good inhibition nature of these derivatives. Also, theoretical calculation supported by other applied corrosion monitoring techniques [33].

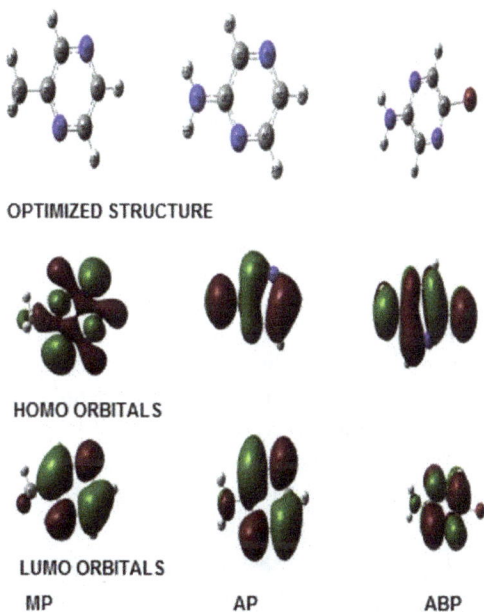

Figure 5. DFT calculation of neutral structure of MP, AP and ABP [Copyright 2014@ Elsevier].

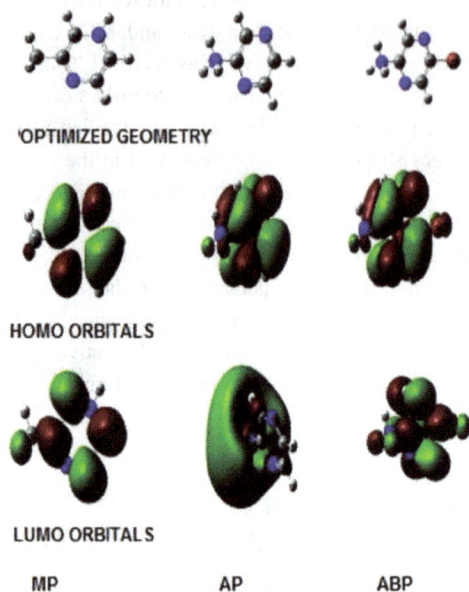

Figure 6. DFT calculation of protonated structure of MP, AP and ABP [Copyright 2014@ Elsevier].

Further aspects

Corrosion inhibition on pyrazine-based derivatives suggests that most of the studies have been on mild steel / steel and aluminium metal. Mainly HCl, H_2SO_4 and NaCl have been frequently used as electrolytic media. But there has been almost no research on metal like Cu, Zn, Mg, Sn, Pb and their metal alloys as a corrosion inhibitor of pyrazine, on which research is yet to be done. Electrolytic medium has been used in low concentration. While the behaviours of corrosion inhibitor in high concentration has been studied in few papers. Also, other media other than NaCl can be used for basic medium. In this way, there is still research into pyrazine corrosion inhibitors is to be done.

Conclusion

The characteristic properties of pyrazine and its derivatives such as solubility, biodegradability, eco-friendly and chemical behaviour make them excellent corrosion inhibitor. Especially the nitrogen atom and π electrons present in them which easily share their electrons onto metal and deposit like a protective sheet over them. In addition to corrosion, the use of pyrazine derivative as a food additive shows its zero-toxic nature. In the past decades, corrosion monitoring techniques such as electrochemical and gravimetric analysis have been used mostly for pyrazine-based corrosion inhibitor. Acidic solutions and basic solutions are commonly used as electrolytic media. In some of the current published work, computation studies such as MD simulations, DFT calculation has also been used as supporting tools with electro chemical analysis. The biggest advantage of theoretical calculus is that it is capable of almost accurately assessing the chemical behaviours of organic molecules along with being less time consuming and chemical free research. Ultimately, research on pyrazine leads to the conclusion that it is a good corrosion inhibitor, those can be easily used.

Abbreviations

WL = Weight loss
EIS = Electrochemical impedance spectroscopy
PDP = Potentiodynamic polarization
OCP = Open circuit potential
SEM = Scanning electron microscopy
EDS = Electron dispersion x-ray spectroscopy
QCC = Quantum chemical calculation
DFT = Density function theory
XPES = X-ray photoelectron spectroscopy
MDS = Molecular dynamic simulation
XRD = X-ray diffraction
FTIR = Furrier transform infrared
AFM = Atomic force microscopy
EDXA = Energy dispersive x-ray analysis
GCMS = Gas chromatography mass spectrophotometer
HE = Hydrogen evolution

Acknowledgement

Authors greatly acknowledge the Principal Govt. Digvijay Autonomous College, Rajnandgaon for providing basic facilities.

References

[1] K. Saha, S. Dutta, A. Ghosh, P. Sukul, D. Banerjee, Adsorption and corrosion inhibition effect of schiff base molecules on the mild steel surface in 1 m hcl medium: A combined experimental and theoretical approach, Phys. Chem. Chem. Phys, 17 (2015) 5679–5690. https://doi.org/10.1039/C4CP05614K

[2]   D.K. Verma, E.E. Ebenso, M.A. Quraishi, C. Verma, Gravimetric, electrochemical surface and density functional theory study of acetohydroxamic and benzohydroxamic acids as corrosion inhibitors for copper in 1M HCl, Results Phys. 13 (2019) 102194. https://doi.org/10.1016/j.rinp.2019.102194

[3]   Y. Qiang, S. Zhang, S. Xu, W. Li, Experimental and theoretical studies on the corrosion inhibition of copper by two indazole derivatives in 3.0% NaCl solution, J. Colloid Interface Sci. 472 (2016) 52–59. https://doi.org/10.1016/j.jcis.2016.03.023

[4]   C. Verma, L. Olasunkanmi, I. Obot, E.E. Ebenso, M.A. Quraishi, 5-Arylpyrimido-[4, 5-b]quinoline-diones as new and sustainable corrosion inhibitors for mild steel in 1 M HCl: a combined experimental and theoretical approach, RSC Adv. 6 (2016) 15639–15654. https://doi.org/10.1039/C5RA27417F

[5]   S. Deng, X. Li, X. Xie, Hydroxymethyl urea and 1, 3-bis (hydroxymethyl) urea as corrosion inhibitors for steel in HCl solution, Corros. Sci. 80 (2014) 276–289. https://doi.org/10.1016/j.corsci.2013.11.041

[6]   D.K. Verma, F. Khan, Corrosion inhibition of mild steel in hydrochloric acid using extract of glycine max leaves, Res. Chem. Intermediat. 42 (2016) 3489–3506. https://doi.org/10.1007/s11164-015-2227-7

[7]   F. Bentiss, M. Traisnel, L. Gengembre, M. Lagrenée, Inhibition of acidic corrosion of mild steel by 3,5-diphenyl-4H-1,2,4-triazole, Appl. Surf. Sci. 161 (2000) 194–202. https://doi.org/10.1016/S0169-4332(00)00287-7

[8]   G. Moretti, 5-Amino- and 5-chloro-indole as mild steel corrosion inhibitors in 1 N sulphuric acid, Electrochim. Acta. 41 (1996) 1971–1980. https://doi.org/10.1016/0013-4686(95)00485-8

[9]   M. Düdükcü, B. Yazici, M. Erbil, The effect of indole on the corrosion behaviour of stainless steel, Mater. Chem. Phys. 87 (2004) 138–141. https://doi.org/10.1016/j.matchemphys.2004.05.043

[10]   W.W.K.R. Mederski, D. Kux, M. Knoth, J. Markus, Pyrido[3,4-b] pyrazines: A new application of 2-Chloro-3,4-diaminopyridine, Heterocycles. 60 (2003) 925 932. https://doi.org/10.3987/COM-02-9666

[11]   M. Bouklaha, A. Attayibatb, B. Hammoutia, A. Ramdanib, S. Radib, M. Benkaddoura , Pyridine–pyrazole compound as inhibitor for steel in 1 M HCl, Appl. Surf. Sci. 240 (2005) 341–348. https://doi.org/10.1016/j.apsusc.2004.07.001

[12] S. Bourichi, Y. Kandri Rodi, M. El-Azzouzi, Y. Kharbach, F. Ouazzani Chahdi, A. Aouniti, nhibitive effect of new synthetized imidazopyridine derivatives for the mild steel corrosion in Hydrochloric acid medium, J. Mater. Environ. Sci. 8(5) (2017) 1696-1807.

[13] A. Ehsani, M.G. Mahjani, R. Moshrefi, H. Mostaanzadeha, J.S. Shayeh, Electrochemical and DFT Study on the Inhibition of 316L Stainless Steel Corrosion in Acidic Medium by 1-(4-nitrophenyl)-5-amino-1H-tetrazole. RSC Adv. 4 (2014) 20031–20037. https://doi.org/10.1039/C4RA01029A

[14] F. El-Hajjaji, B. Zerga, M. Sfaira, M. Taleb, M. Ebn-Touhami, B. Hammouti, S. Al- Salem, H. Benzeid, El. M. Essassi,. Comparative Study of Novel N-Substituted Quinoxaline Derivatives towards Mild Steel Corrosion in Hydrochloric Acid, J. Mater. Environ. Sci. 5(1) (2014) 255–262.

[15] A. Zarrouk, B. Hammouti, A. Dafali, M. Bouachrine, H. Zarrok, S. Boukhris, S.S.A. Al- Deyab, Theoretical Study on the Inhibition Efficiencies of Some Quinoxalines as Corrosion Inhibitors of Copper in Nitric Acid. J. Saudi Chem. Soc. 18 (2014) 450–455. https://doi.org/10.1016/j.jscs.2011.09.011

[16] A.Y. Musa, R.T.T. Jalgham, A.B. Mohamad, Molecular dynamic and quantum chemical calculations for phthalazine derivatives as corrosion inhibitors of mild steel in 1 M HCl, Corros. Sci. 56 (2012) 176–183. https://doi.org/10.1016/j.corsci.2011.12.005

[17] A. Khadiria, R. Saddik , K. Bekkouche, A. Aouniti, B. Hammouti, N. Benchat, M. Bouachrine, R. Solmaz,, Gravimetric, electrochemical and quantum chemical studies of some pyridazine derivatives as corrosion inhibitors for mild steel in 1M HCl solution, J. Taiwan Ins. Chem. Eng. (2015) 1–13. https://doi.org/10.1016/j.jtice.2015.06.031

[18] A. Ghazoui, N. Bencaht, S.S. Al-Deyab, A. Zarrouk, B. Hammouti, M. Ramdani, M. Guenbour, An investigation of two novel pyridazine derivatives as corrosion inhibitor for C38 steel in 1.0 M HCl, Int. J. Electrochem. Sci. 8 (2013) 2272 – 2292.

[19] L. Herrag, A. Chetouani, S. Elkadiri, B. Hammouti, A. Aouniti, Pyrazole derivatives as corrosion inhibitors for steel in hydrochloric acid, Port. Electrochim. Acta. 26 (2008) 211-220. https://doi.org/10.4152/pea.200802211

[20] L. Madhavi, V. Sadasivam, B. Sivasankar, A highly selective synthesis of pyrazine from ethylenediamine on copper oxide/copper chromite catalysts, Catal. Comm. 8 (2007) 1070–1073. https://doi.org/10.1016/j.catcom.2006.06.007

[21] P. Ghosh, A. Mandal, Greener approach toward one pot route to pyrazine synthesis, Green Chem. Lett. Rev. 5 (2012) 127-134. https://doi.org/10.1080/17518253.2011.585182

[22] H. Masuda, S. Mihara, Olfactive properties of Alkylpyrazines and 3-substituted 2-Alkylpyrazines, J. Agr. Food. Chem. 36 (1988) 583-587. https://doi.org/10.1021/jf00081a044

[23] H. Masuda, M. Yoshida, T. Shibamoto, Synthesis of new pyrazines for flavor use, J. Agr. Food Chem. 29 (1981) 944-947. https://doi.org/10.1021/jf00107a014

[24] J. James. Kaminski, D.G. Perkins, J.D. Frantz, D.M. Solomon, A.J. Elliott, P.J.S. Chiu, J.F. Long, Antiulcer Agents. 3. Structure- Activity-Toxicit y Relations hips of Substituted Imidazo[1,243]pyridines and a Related Imidazo[1,2-a Ipyrazine, J. Med. Chem. 30 (1987) 2047-2051. https://doi.org/10.1021/jm00394a019

[25] I.B. Obot, S.A. Umoren, N.K. Ankah, Pyrazine derivatives as green oil field corrosion inhibitors for steel. J. Mol. Liq. 277 (2019) 749-761. https://doi.org/10.1016/j.molliq.2018.12.108

[26] H. Behzadi, S. Manzetti, M. Dargahid, P. Roonasia, Z. khalilniaa, Application of calculated NMR parameters, aromaticity indices and wavefunction properties for evaluation of corrosion inhibition efficiency of pyrazine inhibitors, J. Mol. Struct. 115 (2018) 34-40. https://doi.org/10.1016/j.molstruc.2017.09.029

[27] H. Behzadi, P. Roonasi, M.J. Momeni, S. Manzetti, M.D. Esrafili, I.B. Obot, M. Yousefvand, S. Morteza, M. Khoshdel, A DFT study of pyrazine derivatives and their Fe complexes in corrosion inhibition process, J. Mol. Struct. 1086 (2015) 64–72. https://doi.org/10.1016/j.molstruc.2015.01.008

[28] L.O. Olasunkanmi, Mabine F.S., E.E. Eno, Influence of 6-phenyl-3(2H)-pyridazinone and 3-chloro-6-phenylpyrazine on mild steel corrosion in 0.5 M HCl medium: Experimental and theoretical studies, J. Mol. Struct.1149 (2017) 549-559. https://doi.org/10.1016/j.molstruc.2017.08.018

[29] D. Huang, Y. Tu, G. Song, X. Guo, Inhibition effects of pyrazine and piperazine on the corrosion of Mg-10Gd-3Y-0.5Zr alloy in an ethylene glycol solution, Am. J. Anal. Chem. 3 (2013) 36-38. https://doi.org/10.4236/ajac.2013.46A005

[30] J. Saranya, P. Sounthari, K. Parameswari, S. Chitra, Acenaphtho[1,2-b]quinoxaline and acenaphtho[1,2-b]pyrazine as corrosion inhibitors for mild steel in acid medium, J. Meas. 77 (2016) 175–186. https://doi.org/10.1016/j.measurement.2015.09.008

[31] M. Bouklaha, A. Attayibat, S. Kertit, A. Ramdani, B. Hammouti, A pyrazine derivative as corrosion inhibitor for steel in sulphuric acid solution, Appl. Surf. Sci. 242 (2005) 399–406. https://doi.org/10.1016/j.apsusc.2004.09.005

[32] X. Li, S. Deng, H. Fu, Three pyrazine derivatives as corrosion inhibitors for steel in 1.0 M H_2SO_4 solution, Corro. Sci. 53 (2011) 3241–3247. https://doi.org/10.1016/j.corsci.2011.05.068

[33] I.B. Obot, Z.M. Gasem, Theoretical evaluation of corrosion inhibition performance of some pyrazine derivatives, Corros. Sci. 83 (2014) 359–366. https://doi.org/10.1016/j.corsci.2014.03.008

[34] S. Deng, X. Li, H. Fu, Two pyrazine derivatives as inhibitors of the cold rolled steel corrosion in hydrochloric acid solution, Corros. Sci. 53 (2011) 822–828. https://doi.org/10.1016/j.corsci.2010.11.019

[35] M. Kissi, M. Bouklah, B. Hammouti, M. Benkaddour, Establishment of equivalent circuits from electrochemical impedance spectroscopy study of corrosion inhibition of steel by pyrazine in sulphuric acidic solution, Appl. Surf. Sci. 252 (2006) 4190–4197. https://doi.org/10.1016/j.apsusc.2005.06.035

[36] R. Chopra, K. Kansal, R. Kumar, G. Singh, Electrochemical, morphological and anti-corrosive characteristics of pyrazine derivatives for mild steel corrosion in aggressive medium: A comparative study, J. Fail. Anal. Prev.

[37] S. Zhang, H. Li, L. Wang, D. Liu, E. Ping, P. Zou, T. Ma, N. Li, New pyrazine derivatives as efficient inhibitors on mild steel corrosion in hydrochloric medium, Chem. Eng. Trans. 55 (2016) 289-29.

[38] S.K. Saha, A. Hens, A.R. Chowdhury, A.K. Lohar, N.C. Murmu, P. Banerjee, Molecular Dynamics and Density Functional Theory Study on Corrosion Inhibitory

Action of Three Substituted Pyrazine Derivatives on Steel Surface, 2 (2014) 489-503. https://doi.org/10.13179/canchemtrans.2014.02.04.0137

[39] L.O. Olasunkanmi, Mabine F.S., E.E. Eno, Influence of 6-phenyl-3(2H)-pyridazinone and 3-chloro-6-phenylpyrazine on mild steel corrosion in 0.5 M HCl medium: Experimental and theoretical studies, J. Mol. Struct.1149 (2017) 549-559. https://doi.org/10.1016/j.molstruc.2017.08.018

[40] A.A. Massoud, A. Hefnawy, V. Langer, M.A. Khatab, L. Öhrstrom, M.A.M. Abu-Youssef, Synthesis, X-ray structure and anti-corrosion activity of two silver(I) pyrazino complexes, Polyhedron 28 (2009) 2794–2802. https://doi.org/10.1016/j.poly.2009.05.064

[41] M. Kissi, M. Bouklah, B. Hammouti, M. Benkaddour, Establishment of equivalent circuits from electrochemical impedance spectroscopy study of corrosion inhibition of steel by pyrazine in sulphuric acidic solution, Appl. Surf. Sci. 252 (2006) 4190–4197. https://doi.org/10.1016/j.apsusc.2005.06.035

[42] A.A. Farag T.A. Ali, The enhancing of 2-pyrazinecarboxamide inhibition effect on the acid corrosion of carbon steel in presence of iodide ions, J. Ind. Eng. Chem. 21 (2015) 627-634. https://doi.org/10.1016/j.jiec.2014.03.030

[43] M.Y. Hjouji, M. Djedid, H. Elmsellem, Y.K. Rodi, Y. Ouzidan, F. Ouazzani Chahdi, N.K. Sebbar, E.M. Essassi, I. Abdel-Rahman, B. Hammouti, Corrosion inhibition of mild steel in hydrochloric acid solution by pyrido[2,3-b]pyrazine derivative: Electrochemical and theoretical evaluation, J. Mater. Environ. Sci. 7 (4) (2016) 1425-1435

[44] J.A.M. Abdulwahed, A. Attia, M.R. Elsayad, A.M. Eldesoky, Inhibitive effect of azine and diazine derivatives on the corrosion of cyclic stressed 316l SS in acidic media, Int. J. Sci. Eng. Res. 5 (2014) 342-356.

[45] L. Olasunkanmi, I.B. Obot, M.M. Kabanda, E.E. Ebenso, Some quinoxalin-6-yl derivatives as corrosion inhibitors for mild steel in hydrochloric acid: Experimental and theoretical studies, J. Phys. Chem. C 2015 (1-36). https://doi.org/10.1021/acs.jpcc.5b03285

[46] L. Lei, F. Pan, J. Li, Environmentally friendly corrosion inhibitors for magnesium alloys, in: F. Czerwinski (Ed.), Magnesium Alloys - Corrosion and Surface Treatments, Intech Publishers, 2011, pp. 47-64

[47] H. Lgaz, R. Salghi, S. Jodeh, Y. Ramli, M. Larouj, K. Toumiat, M.A. Quraishi, H. Oudda, W. Jodeh, Understanding the adsorption of quinoxaline derivatives as corrosion inhibitors for mild steel in acidic medium: Experimental, theoretical and molecular dynamic simulation studies, J. Steel. Struct. Constr. 2 (2016) 1-17. https://doi.org/10.4172/2472-0437.1000111

[48] M.Y. Hjouji, M. Djedid, H. Elmsellem, Y..K Rodi, M. Benalia, H. Steli, Y. Ouzidan, F.O. Chahdi, E.M. Essassi, B. Hammouti, Synthesis of novel pyrido[2,3-b]pyrazine derivative evaluated theoretically and electrochemically as a corrosion inhibitor for mild steel in 1M HCl solutions, Der. Pharma. Chemica. 8 (2016) 85-95.

[49] P.M. Nouri, M.M. Attar, Experimental and quantum chemical studies on corrosion inhibition performance of fluconazole in hydrochloric acid solution, Bull. Mater. Sci. 38 (2015) 499–509. https://doi.org/10.1007/s12034-015-0865-4

[50] P. Premkumar, K. Kannan, M. Natesan. Effect of menthol coated craft paper on corrosion of copper in HCl environment, Bull. Mater. Sci. 33 (2010) 307–311. https://doi.org/10.1007/s12034-010-0047-3

[51] M.A. Migahed, M. ABD-EL-Raouf, A.M. AL-Sabagh, H.M. ABD-El-Bary. Corrosion inhibition of carbon steel in acid chloride solution using ethoxylated fatty alkyl amine surfactants. J. Appl. Electrochem. 36 (2006) 395–402. https://doi.org/10.1007/s10800-005-9094-7

[52] R. Mahmoud, N. El-Dn. E.A. Khamis, Corrosion inhibition efficiency, electrochemical and quantum chemical studies of some new nonionic surfactants for carbon steel in acidic media, J. Surfact. Deterg. 17 (2014) 795–805. https://doi.org/10.1007/s11743-014-1565-6

[53] M. Sobhi, R. El-Sayed, M. Abdallah. The effect of non ionic surfactants containing triazole, thiadiazole and oxadiazole as inhibitors of the corrosion of carbon steel in 1M hydrochloric acid, J. Surfact. Deterg. 16 (2013) 937–946. https://doi.org/10.1007/s11743-013-1468-y

[54] U.F. Ekanem, S.A. Umoren, I. Udousoro, A.P. Udoh, Inhibition of mild steel corrosion in HCl using pineapple leaves (ananas comosus l.) extract, J. Mater. Sci. 45 (2010) 5558–5566. https://doi.org/10.1007/s10853-010-4617-y

[55] A. Ousslim, K. Bekkouch, B. Hammouti, A. Elidrissi, A. Aouniti, Piperazine derivatives as inhibitors of the corrosion of mild steel in 3.9 M HCl, J. Appl. Electrochem.39 (2009) 1075–1079. https://doi.org/10.1007/s10800-008-9759-0

[56] I.A. Mohamed, Eco friendly corrosion inhibitors: inhibitive action of quinine for corrosion of low carbon steel in 1 M HCl, J. Appl. Electrochem. 36 (2006) 1163–1168. https://doi.org/10.1007/s10800-006-9204-1

[57] S.A. Umoren, I.B. Obot, N.O. Obi-egbedi, Raphia hookeri gum as a potential eco-friendly inhibitor for mild steel in sulfuric acid. J. Mater. Sci. 44 (2009) 274–279. https://doi.org/10.1007/s10853-008-3045-8

[58] A.A Nazeer, K. Shalabi, A.S. Fouda, Corrosion inhibition of carbon steel by roselle extract in hydrochloric acid solution: Electrochemical and surface study, Res. Chem. Intermed. 41 (2015) 4833–4850. https://doi.org/10.1007/s11164-014-1570-4

[59] Verajeswari. K. Devarayan, G. Mayakrishnan, V. Periasamy, Inhibition of cast iron corrosion in acid, base, and neutral media using schiff base derivatives. J. Surfact. Deterg. 16 (2013) 571–580. https://doi.org/10.1007/s11743-013-1439-3

[60] D.K. Verma, F. Khan, C.B. Verma, R. Susai, M.A. Quraishi, Experimental and theoretical studies on mild steel corrosion inhibition by the grieseofulvin in 1M HCl, Eur. Chem. Bull. 6 (2017) 21-30. https://doi.org/10.17628/ecb.2017.6.21-30

[61] G. Mohamed, B. Ahmad, Z. Basem, Inhibition of mild steel corrosion in sulfuric acid solution using collagen. Res. Chem Intermed. 41 (2015) 7245-7261. https://doi.org/10.1007/s11164-014-1809-0

[62] P. Premkumar, K. Kannan, M. Natesan, Effect of menthol coated craft paper on corrosion of copper in HCl environment. Bull. Mater. Sci. 33 (2010) 307–311. https://doi.org/10.1007/s12034-010-0047-3

[63] V. Kumar, K.P.S.N. Pillai, M.R. Thusnavis G., Green corrosion inhibitor from seed extract of areca catechu for mild steel in hydrochloric acid medium, J. Mater. Sci. 46 (2011) 5208–5215. https://doi.org/10.1007/s10853-011-5457-0

[64] W.H. Li, H. Qiao, T.Z. Sheng, L.P. Chang, R.H. Bao, Some new triazole derivatives as inhibitors for mild steel corrosion in acidic medium, J. Appl. Electro. Chem. 38 (2008) 289–295. https://doi.org/10.1007/s10800-007-9437-7

Theory and Applications of Green Corrosion Inhibitors   Materials Research Forum LLC
Materials Research Foundations **86** (2021) 183-203   https://doi.org/10.21741/9781644901052-7

Chapter 7

Biological Corrosion Inhibitors for Concrete

M. Kanwal[1], R. Bagheri*[2,3], A.G. Wattoo*[3,4], M.I. Irshad[4], R.A. Khushnood*[1], Z. Song[3]

[1] NUST Institute of Civil Engineering (NICE), National University of Sciences and Technology (NUST), Islamabad-44000, Pakistan

[2] School of Physical Science and Technology, College of Energy, Soochow Institute for Energy and Materials Innovations and Key Laboratory of Advanced Carbon Materials and Wearable Energy Technologies of Jiangsu Province, Soochow University, Suzhou 215006, China

[3] CAS Key Laboratory of Magnetic Materials and Devices, Zhejiang Province Key Laboratory of Magnetic Materials and Application Technology, Ningbo Institute of Materials Technology and Engineering, Chinese Academy of Sciences, Ningbo-315201, China

[4] Department of Physics, Khwaja Fareed University of Engineering and Information Technology, Rahim Yar Khan-64200, Pakistan

* arsalan.khushnood@nice.nust.edu.pk, dragwattoo@gmail.com, bagheri.robabeh@gmail.com

Abstract

Corrosion in the reinforced concrete structures is a promising concern for material and structural engineers. Chemical corrosion inhibitors are usually suggested by the corrosion specialists in this regard, which adds an enormous amount of cost to the total materials charges and some are harmful to life forms. Therefore, some green corrosion inhibitors are introduced in this study which are relatively cheaper and holds good compatibility with concrete. We have highlighted the mechanisms of protection, advantages and disadvantages of different green biological corrosion inhibitors by broadly categorizing them into two types. This classification is based on the origin and mechanism of corrosion inhibition of the respective class. The first class is microbial inhibitors which prevent the onset of corrosion by making biofilms and precipitating calcite on steel surface and concrete voids respectively. And the second class is of botanical inhibitors which are the extracts of plants and protect the steel bars by forming biofilms. So both of the biological inhibitors are efficient to eliminate the corrosion of reinforced concrete structures.

Keywords

Green Corrosion Inhibitors, Concrete, Biological Inhibitors, Biofilm, Calcite

Contents

1.   Introduction

Concrete is the most abundantly used construction material worldwide due to its copiousness, lower cost and superior mechanical properties [1,2]. Globally annual production of concrete is more than 3.8 billion cubic meter [3]. Besides this, it is weak in tension due to its fragility, when loaded in bending. Reinforced concrete (RC) having steel bars is the best remedy to give durability and strength in both aspects. But during the service life undergo heavy loads, settlements, freeze and thaw actions, which cause cracking in concrete. These cracks and interconnecting pores give paths to chlorides, moisture, oxygen (O_2) and carbon dioxide (CO_2) to penetrate into the concrete. The penetration of such aggressive substances not only reduces the alkalinity of concrete by reducing its pH but also causes the degradation of passive layer on steel bars and corrode them [4,5]. If these cracks are not repaired at proper time, ultimately lead to a structural failure. Because rust on steel bars expand with time and inserts an internal pressure on the concrete.

To mitigate corrosion of RC members various organic, inorganic and green corrosion inhibitors have been explored. But this chapter is focused on the green biological corrosion inhibitors for RC members shown in Fig. 1.

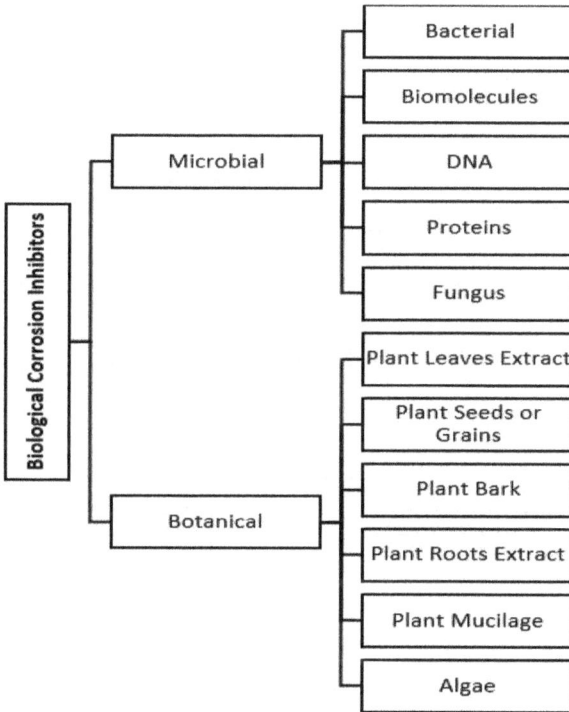

Figure 1.   Classification of biological corrosion inhibitors

2.   Biological Corrosion Inhibitors

2.1   Microbial

Microbial inhibitors are those where we use some microorganisms like bacteria, fungus or biomolecules *etc.* for the instant repair of micro cracks which are the crucial pathways for chlorides and other violent substances.

2.1.1   Bacterial

Various bacteria have the ability to precipitate $CaCO_3$ in the interconnecting pores and cracks of the concrete and block the passage of aggressive species. Different bacterial species precipitate $CaCO_3$ differently in concrete i.e. non-ureolytically, ureolytically or metabolically. Fig. 2 is showing the two types of bacterial solution in first type bacteria can be added directly after 24 hours of incubation in the broth media and they can also be added in the form of bacterial spores' solution, bacterial spores survive for longer period in concrete compared to the direct bacterial solution.

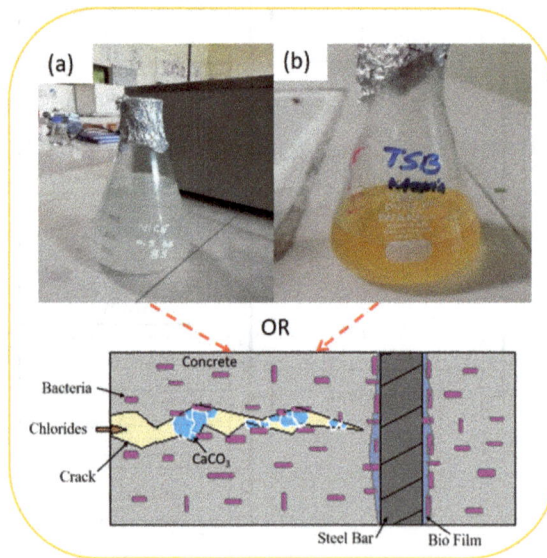

Figure 2.   Bacterial solution to be used in concrete and schematic representing of the calcite precipitation and formation of passive biofilm by bacteria (a) Bacterial spores' solution (b) Direct bacterial solution

2.1.1.1 Ureolytic

Ureolytic bacteria decompose urea by the help of their urease enzymes. Examples of ureolytic bacteria are *Bacillus sphaericus*, *Bacillus pasteurii*, *Bacillus psychrophilus*, *Bacillus globisporus*, *Planococcus okeanokoites*, and *Filibacter limicola* etc. As urea $(CO(NH_2)_2)$ is the constituent of bacterial metabolic activities, therefore this process is the cause of production of ammonium and carbonates. As a result, increase the number of carbonates around the bacterial cells and also increase the alkalinity. Further, ammonium (NH_4^+) and carbonic acids (CO_3^{2-}) are produced by the hydrolysis of these constituents. Chemical equations of the whole process for the formation of CO_3^{2-} and NH_4^+ from $CO(NH_2)_2$ are as follows [6].

$$CO(NH_2)_2 + H_2O \rightarrow NH_2COOH + NH_3 \tag{1}$$

$$NH_2COOH + H_2O \rightarrow NH_3 + H_2CO_3 \tag{2}$$

$$H_2CO_3 \rightarrow HCO_3^- + H^+ \tag{3}$$

$$2NH_3 + 2H_2O \rightarrow 2NH_4^+ + 2OH^- \tag{4}$$

$$HCO_3^- + H^+ + 2NH_4^+ + 2OH^- \rightarrow CO_3^{2-} + 2NH_4^+ + 2H_2O \tag{5}$$

The bacterial cell wall having negative charge attracts cations (Ca^{2+}) from the surrounding environment and attach on their cell surface. These Ca^{2+} react with the CO_3^{2-} (from equation 5) for the formation of calcium carbonate or calcite (equation 7) at the surface of bacterial cells.

$$Ca^{2+} + Cell \rightarrow Cell \text{ - } Ca^{2+} \tag{6}$$

$$Cell \text{ - } Ca^{2+} + CO_3^{2-} \rightarrow Cell \text{ - } CaCO_3 \tag{7}$$

Fig. 3 representing the schematic illustration of the biological formation of *CaCO₃* by bacterial activities in concrete to seal the cracks.

$$CO_3^{2-} \quad NH_4^+$$

Hydrolysis of urea

Ureolytic bacteria   Calcium ions $CaCO_3$

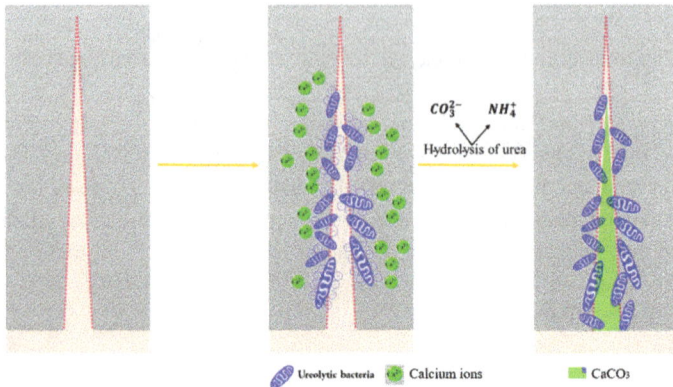

Figure 3.   Schematic diagram representing the calcite precipitation and formation of passive biofilm by bacteria

2.1.1.2 Non-ureolytic

Some bacteria precipitate $CaCO_3$ by their metabolic activities in the presence of a calcium source as bacterial food and oxygen (equation 8).

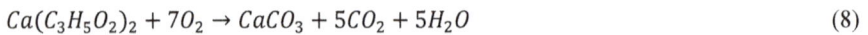

$$Ca(C_3H_5O_2)_2 + 7O_2 \rightarrow CaCO_3 + 5CO_2 + 5H_2O \qquad (8)$$

This produced CO_2 further react with portlandite $[Ca(OH)_2]$, which is the cement hydration product, and accelerate the self-healing process by making added calcite.

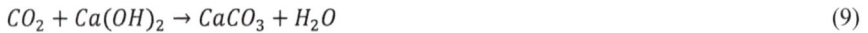

$$CO_2 + Ca(OH)_2 \rightarrow CaCO_3 + H_2O \qquad (9)$$

This biological process of instant cracks repairing without compromising the concretes mechanical properties is called autonomous healing [7,8]. Besides, bacteria can directly protect the steel bars from corrosion by making thick sticky biofilms on steel surface. This protective layer is formed by some bacterial strains when change to sticky biofilm cells in response to the exposure of oxygen, nutrients or both. Various bacterial species form biofilm at the moisture-air interface, as reported for corrosion inhibition of mild and stainless steels [9,10] so that this film inhibit the passage of aggressive species to reach the steel bars in concrete.

Theory and Applications of Green Corrosion Inhibitors   Materials Research Forum LLC
Materials Research Foundations **86** (2021) 183-203   https://doi.org/10.21741/9781644901052-7

Figure 4 (left) is the real case image of calcite precipitated by *Bacillus Subtilis* in about 0.2mm wide concrete crack, measured by crack width measuring microscope. Figure 4 (right) is showing Scanning Electron Microscopy (SEM) micrograph of $CaCO_3$, there are round to spherical shape calcite crystals precipitated by *Bacillus Subtilis* in concrete cracks, in micro to nano size range.

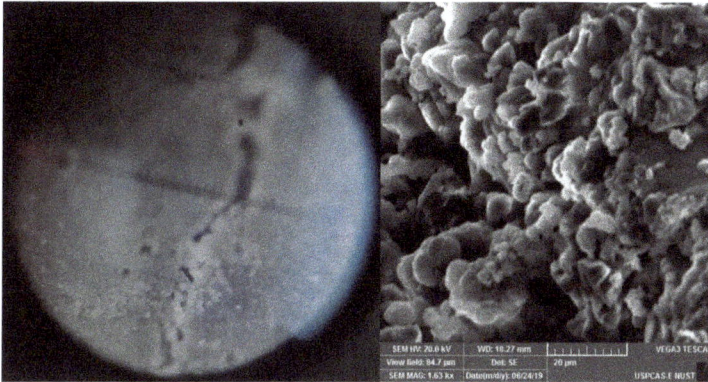

Figure 4.   Production of calcite by bacteria: (left) Cracks closure by bacterial calcite precipitation; (right) SEM image of microbially precipitated calcite in concrete cracks

Wadood et al. [10] demonstrated that *Bacillus Subtilis* and *Pseudomonas Aeruginosa* have the corrosion resistance ability to protect the stainless steel 304 surface from oxidation by the formation of biofilm on the metal surface. Rajasekar et al. [11] also reported that the corrosion potential (E_{corr}) and pitting potential (E_{pit}) were lower in the presence of the previously mentioned bacteria in an organic medium. Salts like nitrates or phosphates can enhance the bacterial metabolites which result in corrosion inhibition. But these studies were conducted on steel metals not on reinforced concrete members. *Bacillus Subtilis* form a stable, more compact, uniform and defect free biofilm compared to many other corrosion inhibiting bacteria that's why they are superior among all bacteria [12].

2.1.2 Nitrate reducing bacteria

Nitrate (NO_3^-) reducing bacteria reduce the NO_3^- by the bacterial oxidation of organic compounds and these NO_3^- then behaves like the electron acceptors (equations 10-13) as

Theory and Applications of Green Corrosion Inhibitors   Materials Research Forum LLC
Materials Research Foundations 86 (2021) 183-203   https://doi.org/10.21741/9781644901052-7

a substitute to oxygen. Then calcite is precipitated in the presence of some calcium ions as shown in equation (14) [13].

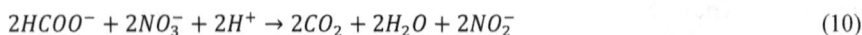

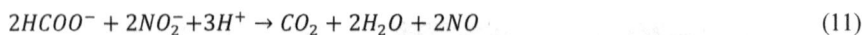

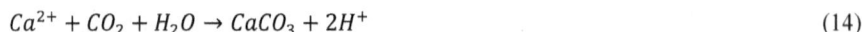

$$2HCOO^- + 2NO_3^- + 2H^+ \rightarrow 2CO_2 + 2H_2O + 2NO_2^- \tag{10}$$

$$2HCOO^- + 2NO_2^- + 3H^+ \rightarrow CO_2 + 2H_2O + 2NO \tag{11}$$

$$2HCOO^- + 2NO + H^+ \rightarrow CO_2 + H_2O + N_2O \tag{12}$$

$$HCOO^- + N_2O + H^+ \rightarrow CO_2 + H_2O + N_2 \tag{13}$$

$$Ca^{2+} + CO_2 + H_2O \rightarrow CaCO_3 + 2H^+ \tag{14}$$

By the reduction of NO_3^-, corrosion of RC members can be reduced through two mechanisms at the same time, one is by blocking the paths of aggressive substances by the precipitation of calcite in the pores and cracks. Secondly, the biochemically production of NO_2^- (which is a well-known corrosion inhibitor [13]) (equation 10) to protect the steel bars from rusting.

Axenic (*Diaphorobacter nitroreducens, Pseudomonas aeruginosa)* and non-axenic (*activated compact denitrifying core*) cultures of nitrate reducing bacteria were used in concrete pore solution to explore the corrosion inhibiting efficiency by calcite precipitation and making biofilms on steel surface [13]. Axenic cultures should be used with some immobilizers to enhance the nitrate reduction but the non-axenic are self-immobilized and give better corrosion protection than the axenic culture. These bacteria can survive for three days in mortar and activate when the pH drops to 10.

2.1.3 Biomolecules

Biomolecules obtained from bacterial cells are the ecofriendly solution for corrosion protection of reinforced concrete members. They form biofilm on steel bars and inhibit corrosion by adding just 1g per litter of water [14].

The schematic diagram of steel surfaces is shown in Figure 5 which is the good visual representation of biofilm formed by the sticky biomolecules. This image also indicates that biomolecules have the tendency to adhere on the surface of steel and can be stared like an excellent inhibitor against the corrosion of steel in chloride ions solution.

Biomolecules can control both anodic and cathodic corrosion reactions by blocking the reaction sites. They create ohmic resistance zones between steel and electrolyte by eliminating the chemical reactions which leads to the formation of corrosion products on metal surface. Simply speaking, these inhibitors slow down the galvanic reactions and results in passivity and increase of alkalinity.

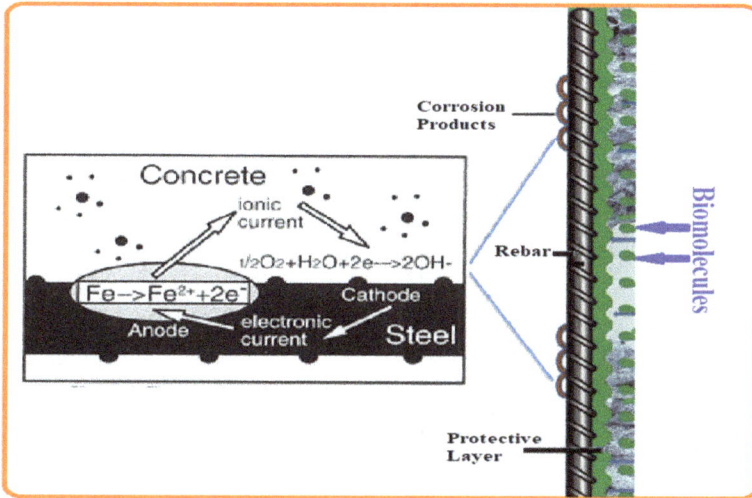

Figure 5.   *Schematic presentation of the mechanism of adhering of biomolecules on steel bars to protect from rusting*

Electrochemical tests indicate that biomolecules reduce both the anodic and cathodic polarization curves thus we can say that biomolecules affect both of the half-cell reactions. This property classifies them as mixed type corrosion inhibitors. Therefore, corrosion protection is achieved due to the formation of barrier on the steel surface which inhibit penetration of chlorides to contact the steel surface. Biomolecules enhance the polarization resistance (Rp) of steel due to the formation of biofilm which also helps in the determination of inhibition efficiency of steel by using equation 15. Biomolecules enhance the corrosion inhibiting efficiency of RC members by 58.6% [14] compared to the blank or control concrete.

$$\eta(\%) = \frac{R_{P_{biomolecule\,Mix}} - R_{P_{Control\,Mix}}}{R_{P_{biomolecule\,Mix}}} \times 100 \tag{15}$$

Where;

η is the corrosion inhibition efficiency of biomolecules containing concrete in percentage, $R_{P_{biomolecule\ Mix}}$ is the polarization resistance of biomolecules containing concrete in Ohms and $R_{P_{Control\ Mix}}$ is the polarization resistance of control mix or blank concrete in Ohms.

2.1.4 Deoxyribonucleic acid (DNA)

DNA also have the ability to mitigate corrosion of RC members. Its molecular structure consists of deoxy-nucleosides, Figure 6(a), where different types of bases (B) may present, like guanine, thymine, adenine or cytosine. Corrosion inhibitors consisting of DNAs are composed of about 20 to 80 oligonucleotides similar to shown in Figure 6(b).

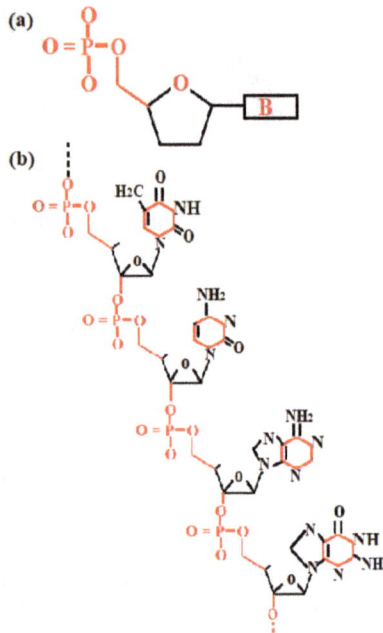

Figure 6.   The molecular structure of DNA (a) deoxy-nucleoside (b) oligonucleotide

Firstly corrosion inhibition properties of DNA were tested for 10 mm size carbon steel in simulated concrete pore (SCP) solution where it showed the admirable inhibition at varying amount of chlorides [15]. It gives best results at DNA's concentration of only 0.0025%. There are two steps in corrosion prevention by DNAs. DNA firstly adsorb on the steel surface and inhibit the cathodic process of corrosion propagation by making a passive film. Then by increasing the chlorides concentration or at next level of corrosion, DNA get adsorbed at active corrosion sites and inhibit the anodic corrosion reaction. Therefore, we can classify this type of inhibitor as mixed type inhibitor.

This biological corrosion inhibitor behaves similar in real cementitious environment as well [16]. Electrochemical studies show that in 3.5% NaCl solution this corrosion inhibitor is superior than phosphate, a commercially available inhibitor, with DNA concentrations of 0.0025%. Because these ecofriendly natural species don't put any toxic or damaging impact on the mechanical properties of concrete.

2.1.5 Mussel adhesive proteins

It is natural glue like adhesive material obtained from the different parts of mussel and called as mussel adhesive proteins (MAP). It can stick on metal surfaces, make films and protect to the steel rebars from corrosion [17–19]. At earlier stages this film has micron to sub-micron sized holes, as anodic area is lesser so enhance the local corrosion. But these anodic sites removed at later stages which indicates the formation of passive layer on the bare steel surface. It can be considered that the decrease in corrosion at later stages is due to the blocking of corrosion pits by rust components but it's not true. This is because depth of holes does not decrease continuously but increases at earlier and then became stable due to passivation. This indicate that decrease of corrosion is due to the self-healing mechanism of MAP with some nanocomposites i.e. CeO_2. This passivation or self-healing is the result of the reaction of iron ions (Fe^{2+}) and MAP in the composite film. Initially, iron dissolution takes place inside the defects in metal surface which results in the release of Fe^{2+} and make complex iron catecholato e.g. $Fe(Ca_3)_3^-$ and adhere with the nanocomposite layer. This iron dissolution results in increase of pit depth initially and later these dissolved ions adhere to the nearby film and start filling the pit and reduction in pit depth takes place. This process results in becoming more compact and passive layer and thus diminish the pitting corrosion. So, the whole process of increase or decrease of pit depth depends on the formation and removal of iron catecholato. But later hole depth become stable and automatically heal the defects in metal by making passive film. Besides, self-healing property of MAP is applicable even for smaller defects in micrometer sizes and provide long term corrosion protection. This self-healing mechanism is schematically illustrated in Figure 7.

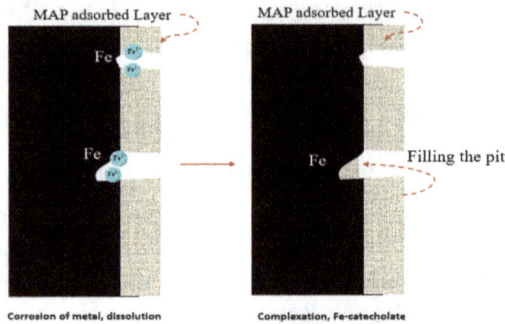

Figure 7.   *Schematic diagram of self-healing mechanism of metal in chlorides environment by MAP/Nanocomposites*

2.1.6 Fungus

Fungi are similar to plants and animals, but it is closer to the insects and microbes because it has chitin similar to microbes instead of having cellulose. Some fungus species like *Trichoderma reesei* (ATCC13631) spores can also be used in concrete to get prevention from corrosion. Because they also have the ability to precipitate calcite in highly alkaline environment like concrete [20]. Calcite fills the cracks and voids and make a more water tight concrete by reducing the permeability of harmful substances from environment so fungi can also be used to protect the RC corrosion after checking the compatibility of fungus with steel.

2.2 Botanical

Botanical or agricultural corrosion inhibitors are the natural green inhibitors produced by the extracts of different parts of different plants or trees. These extracts are added into the concrete during mixing; they not only enhance the corrosion resisting power of the RC but also elevate the mechanical properties in various aspects.

Various types of botanical inhibitors have been used by researchers at material level till now. Azadirachta indica, Aloevera, Anthocleista djalonensis, Rice husks, Vernonia amygadalina, Bambusa arundi- nacea, Cymbopogon citratus, Rhizophora mangle L. bark extract, Chamerops humilis L. extract, Eucalptus globulus, Punica granatum, Olea europaea, Prosopis juliflora, Cactus mucilage, seaweed extract, Rosa damascene leaves, Green bambusa arundinaceous leaves, Ginger Extract, palm oil leaves etcetera are some examples of botanical corrosion inhibitors. Few of these are discussed here in detail.

2.2.1 Extract of tree/plant leaves

Aqueous extracts of different trees or plants can be used as biological or natural corrosion inhibitors of RC members. These extracts are usually prepared by drying and grinding the fresh leaves and fine powder is refluxed to get acidic or ethanol extracts. Corrosion inhibition property or antioxidative nature of leave extracts directly depends upon the content of phenolic compounds in their composition. If any extract has more phenolic compounds, they will have more resistance to corrosion and vice versa. Aqueous extracts contain polar groups of oxygen, nitrogen, phosphorous and sulphur. Organic molecules of these polar groups adhere on the surface of steel bars to make a protective film which prevent it from rusting. Absorption of leave-extract molecules on metal surface reduce the interfacial capacitance of metal and electrolyte compared to the bare steel.

Leaves extract consists of Juliflorine, five-hydroxy-triptamine, isorhamnetin-five-diglucoside etcetera. As we know the most of the familiar corrosion inhibitors contain organic compounds of heteroatoms, triple/conjugated double p-electron and electronegative type functional groups having aromatic groups in their structures are beneficial for making bond with metal as shown in Figure 8. Besides, the presence of other polar groups like oxygen, nitrogen or p-electrons helps them to get absorbed on the metal surface. These molecules can exist in neutral or cationic form in aqueous media.

There might be two methods involved for corrosion inhibition; (1) firstly Cl^- absorb on the metal surface and make it negatively charged like cathode and then electrostatically inhibitor molecules also adhere over the embedded metal surface. (2) Secondly electrons from oxygen, nitrogen and aromatic rings p-electrons of inhibitors absorb on the metal and anodically reduce the corrosion reaction. So plant extract can control both anodic and cathodic reactions and hence so are mixed mode inhibitors.

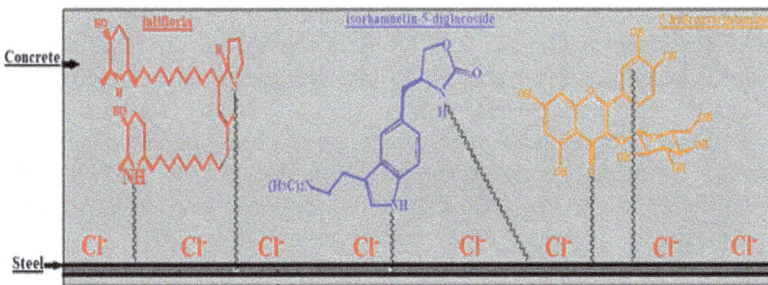

Figure 8.   Schematic representation of inhibitive action of plant extract

Similarly, some other plant/tree leaves which are efficient for inhibiting the RC corrosion are given below;

a) *Eucalptus globulus, Punica granatum, Olea europaea,* have been used at material level in the concrete pore solution and exhibited positive effects on corrosion performance [21]. Leaves extracts of above mentioned trees can be classified as mixed type of corrosion inhibitors because they can positively affect both of the anodic and cathodic corrosion reactions.

b) S. P. Palanisamy used the leaves extract of plant *Prosopis Juliflora* to prevent the reinforced steel bars from rusting [22]. *P. Juliflora* leaves extract react with some free chloride ions, present in the aqueous phase, in the concrete matrix to bound them. Thus, free Cl^- become inactive after bounding with the plant extract and corrosion of RC may be eliminated or reduced. Its composition involves Juliflorine, five-hydroxy-triptamine, isorhamnetin-five-diglucoside etcetera. Therefore, it can be said that *P. Juliflora's* leaves extract can efficiently inhibit the corrosion of RC in chlorides solution but exact amount/concentration of extracted molecules for proper corrosion inhibition is unknown yet.

c) Similarly, leaves extract of *Chamaerops humilis* also has the ability to improve corrosion resistance efficiency by about 60% compared to the blank concrete [23]. Presence of higher amount of natural chemical compounds enables the higher molecular absorption which elevate the polarization resistance of the metal by improving the passivity.

d) *Alovera* and *Neem* (*Azadirachta Indica*) are the most growing plants and similar to many other plants they also have the leaves rich in heteroatoms of sulphur, oxygen, nitrogen and phosphorus [24,25]. They are one of the ecofriendly solutions of corrosion without affecting the strength of the concrete by making a protective film on steel. Neem is more superior if comparing the extracts of both plants in terms of corrosion inhibiting efficiency.

e) Extract from *Anthocleista Djalonensis* leaves has about 97% experimental inhibiting efficiency in RC members. Its molecules absorption on the steel surface follow the Langmuir model [26].

f) *Vernonia Amygdalina* leaves extract also plays a corrosion inhibition role in RC and if comparing to the two commercially available inhibitors (Sodium Nitrite and Calcium Nitrite), *Vernonia Amygdalina* act better than sodium nitrite and approximately similar to calcium nitrite corrosion inhibitor [27].

g) Resistance to charge transfer increases with the increase in the concentration of *Rosa Damascene* leaves extract in RC [28]. As its composition has also involved the alcohols, ketons, phenols or aldehydes polar heteroatoms of oxygen, sulpher, nitrogen and aromatic rings as discussed earlier for other plant extracts. Heteroatoms offer additional sites and make a passive film which inhibit the percolation of depolarizing agents.

h) Fresh leaves of *Bambusa Arundinacea* are used as corrosion mitigators in reinforced concrete when its ecofriendly, non-toxic and hydrophobic extract is added in concrete during casting [29].

i) A tropical plant, *Cymbopogon citratus* or lemon grass extract is superior than sodium nitrite in inhibiting corrosion [30]. A minor quantity of its extract is enough to add compressive strength and corrosion inhibition efficiency of RC.

j) *Elaeis guineensis* also called African oil palm or simply oil palm leaves are beneficial for controlling corrosion. It can give about 95% improved corrosion inhibiting efficiency upon adding with some nano intrusions in reinforced concrete [31].

2.2.2 Bark extract of trees/plants

Bark extract from trees like *Rhizophora Mangle* is one of the best corrosion inhibitors for RC matrix in industrial or microbial environment. It is also prepared by grinding the bark into the fine powder and then acidic or methanolic extract is collected by using a condenser equipped Soxhlet extractor. This extract gives about 78.6% (experimental) efficiency to inhibit rusting in 0.5M H_2SO_4 solution with 0.167% concentration of extract [32].

2.2.3 Seeds or grains

Rice Husk being the waste of the rice is the most economical inhibitor for the construction of corrosion resisting RC members. Like plant leaves extracts rice husk also make protective layer around steel bars and reduce the corrosion current densities when added 3% by weight of cement in fresh concrete during mixing [33].

2.2.4 Plant roots extracts

Similar to other parts extracts, roots of some plants also give the similar results of corrosion inhibition, like ginger. It is the rhizome/roots of *Zingiberaceae* family plant, commonly grow in tropical and subtropical regions and it has a number of advantages because it is economical, non-toxic, readily and widely available [34]. Ginger consists of two types of extracts, volatiles and non-volatiles (phenolic compounds superfluous). These phenolic constituents are prolific in inhibiting corrosion because they elevate the passivity by adhering or absorbing on the metal surface. Ginger extract also has aromatic functional groups, p-electrons and oxygen atoms so have additional source of corrosion resistivity. These atoms form a carbonaceous organic layer on steel surface which act as a barrier for aggressive substances.

Ginger extract is also a mixed type corrosion inhibitor because firstly it forms an absorption layer on original passive oxides film of rebar to restrict the cathodic reaction. But at next stage of corrosion or at higher concentrations of chlorides ginger extract induce passivity by restring the anodic process.

2.2.5 Plants mucilage

Mucilage is the gluey material produced by almost every plant and also by some microorganisms, it consists of glycoprotein and an exopolysaccharide. *Cactus* has rich amount of mucilage compared to other plants [35]. It can be added in RC as an organic material because it contains amino acids to inhibit rusting of rebars. It involves a corrosion inhibition mechanism where chelate substrate is formed on the metal surface by shifting electrons from polysaccharide O^{2-} sites to metal and a covalent bond is formed. Shifting is caused by the presence of organic hydroxyl groups of polysaccharides which deprotonate in alkaline medium and apply a negative charge on polysaccharide chains. Then the cathodic sites attract them to form a thick, stable and sticky film. As polymers strongly stick on metal surfaces compared to monomers that's why they are categorized as good corrosion inhibitors.

Corrosion inhibition efficiency by *cactus mucilage* is directly proportional to its concentration because absorption layer on metal surface thicken with the addition of gluey compound. It consists of polysaccharide which enhance the impedance of rebars against rusting.

2.2.6 Algae

Marine seaweed extract or *Macrocystis Pyrifera* is a form of kelp or algae is efficient to improve the mechanical and durability properties of concrete [35]. It contains multiple compounds to inhibit corrosion of RC like amino acids, oligosaccharides, fucoxanthin and alginate (polysaccharides). It is a mixed (anodic/cathodic) corrosion inhibitor. Seaweeds has histidine, methionine and glutamic acid and these all are as separate corrosion inhibitors therefore, they are capable of forming a thick passive layer on metal surface.

3.   Comparison

- If comparing both of the biological inhibitors, then microbial inhibitors are more efficient than botanical admixtures because microbial species can prevent corrosion by double action. Firstly, by the precipitation of calcium carbonate crystals in the concrete pores. This calcite not only enhance the mechanical properties of the concrete matrix but also reduce the sorptivity or permeability of the aggressive substances. Secondly, some microbes can efficiently make a thick, defect free passive layer on the embedded steel bars. And microbial spores can survive for longer periods in dormant stage and activate only when needed, means they activate upon exposure of oxygen or aggressive environments upon the production of minor cracks in the concrete during its service life.

- But botanical extract molecules only adhere on the metal surface and make well compacted passive films for corrosion inhibition. They do not fill the concrete voids and cracks. The thickness and long lasting properties of these botanical films are yet to be explored.

- Plant extracts can be added directly into the concrete during the mixing and there is no chance of getting defected by the harsh concrete mixing and no threat of reduction in concentration of molecules if adding directly. But microbial species need extra protection in harsh environments so they must be added with some immobilizing material to prevent their concentration or life.

Conclusion

This chapter explored the two major biological inhibitors which are microbial and botanical. The study shows that both inhibitors are reliable, cheaper and environment friendly. The biological inhibitors in different forms can be introduced into the concrete during mixing which reduces the permeability of concrete by producing calcite healing products into the interconnecting pores and sometimes swabbed over the reinforcing steel. Both inhibitors can efficiently inhibit the corrosion of steel bars by the formation of sticky film on the surface of bars, hinder the chemicals to react with steel, and thus preventing the corrosion.

References

[1]   N. Roghanian, N. Banthia, Development of a sustainable coating and repair material to prevent bio-corrosion in concrete sewer and waste-water pipes, Cem. Concr. Compos. 100 (2019) 99–107. https://doi.org/10.1016/j.cemconcomp.2019.03.026

[2]   C. Meyer, Concrete Materials and sustainable development in the united states, 83 Concr. Mater. Sustain. Dev. (1992) 1–10.

[3]   S. Mindess, Introduction, Dev. Formul. Reinf. Concr. (2008) xiii–xiv.

https://doi.org/10.1016/B978-1-84569-263-6.50017-1

[4]   Y. Wang, Y. Zuo, Y. Tang, Inhibition effect and mechanism of sodium oleate on passivation and pitting corrosion of steel in simulated concrete pore solution, Constr. Build. Mater. 167 (2018) 197–204. https://doi.org/10.1016/j.conbuildmat.2018.01.170

[5]   J.K. Das, B. Pradhan, Effect of cation type of chloride salts on corrosion behaviour of steel in concrete powder electrolyte solution in the presence of corrosion inhibitors, Constr. Build. Mater. 208 (2019) 175–191. https://doi.org/10.1016/j.conbuildmat.2019.02.153

[6]   K. Vijay, M. Murmu, S. V. Deo, Bacteria based self healing concrete – A review, Constr. Build. Mater. 152 (2017) 1008–1014. https://doi.org/10.1016/j.conbuildmat.2017.07.040

[7]   E. Schlangen, H. Jonkers, S. Qian, A. Garcia, E. Schlangen, H. Jonkers, S. Qian, A. Garcia, Fract. Mech. Concr. Concr. Struct. Adv. Fract. Mech. Concr. Oh, Al. (2010) 291–298.

[8]   D. Gat, M. Tsesarsky, D. Shamir, Z. Ronen, Accelerated microbial-induced $CaCO_3$ precipitation in a defined coculture of ureolytic and non-ureolytic bacteria, Biogeosciences. 11 (2014) 2561–2569. https://doi.org/10.5194/bg-11-2561-2014

[9]   M. Morikawa, Beneficial biofilm formation by industrial bacteria Bacillus subtilis and related species, J. Biosci. Bioeng. 101 (2006) 1–8. https://doi.org/10.1263/jbb.101.1

[10]   H.Z. Wadood, A. Rajasekar, Y.P. Ting, A.N. Sabari, Role of Bacillus subtilis and Pseudomonas aeruginosa on Corrosion Behaviour of Stainless steel, Arab. J. Sci. Eng. 40 (2015) 1825–1836. https://doi.org/10.1007/s13369-015-1590-4

[11]   A. Rajasekar, Y.P. Ting, Role of inorganic and organic medium in the corrosion behavior of bacillus megaterium and pseudomonas sp. in stainless steel SS 304, Ind. Eng. Chem. Res. 50 (2011) 12534–12541. https://doi.org/10.1021/ie200602a

[12]   Z. Guo, T. Liu, Y.F. Cheng, N. Guo, Y. Yin, Adhesion of Bacillus subtilis and Pseudoalteromonas lipolytica to steel in a seawater environment and their effects on corrosion, Colloids Surfaces B Biointerfaces. 157 (2017) 157–165. https://doi.org/10.1016/j.colsurfb.2017.05.045

[13]   Y.Ç. Erşan, H. Verbruggen, I. De Graeve, W. Verstraete, N. De Belie, N. Boon, Nitrate reducing $CaCO_3$ precipitating bacteria survive in mortar and inhibit steel corrosion, Cem. Concr. Res. 83 (2016) 19–30.

https://doi.org/10.1016/j.cemconres.2016.01.009

[14] V. Shubina, L. Gaillet, T. Chaussadent, T. Meylheuc, J. Creus, Biomolecules as a sustainable protection against corrosion of reinforced carbon steel in concrete, J. Clean. Prod. 112 (2016) 666–671. https://doi.org/10.1016/j.jclepro.2015.07.124

[15] S. Jiang, L. Jiang, Z. Wang, M. Jin, S. Bai, S. Song, X. Yan, Deoxyribonucleic acid as an inhibitor for chloride-induced corrosion of reinforcing steel in simulated concrete pore solutions, Constr. Build. Mater. 150 (2017) 238–247. https://doi.org/10.1016/j.conbuildmat.2017.05.157

[16] S. Jiang, S. Gao, L. Jiang, M.-Z. Guo, Y. Jiang, C. Chen, M. Jin, S. Bai, Effects of Deoxyribonucleic acid on cement paste properties and chloride-induced corrosion of reinforcing steel in cement mortars, Cem. Concr. Compos. 91 (2018) 87–96. https://doi.org/10.1016/j.cemconcomp.2018.05.002

[17] C. Chen, F. Zhang, C. Lin, J. Pan, Corrosion Protection and Self-Healing of a Nanocomposite Film of Mussel Adhesive Protein and CeO$_2$ Nanoparticles on Carbon Steel , J. Electrochem. Soc. 163 (2016) C545–C552. https://doi.org/10.1149/2.0521609jes

[18] C. Chen, R. Hou, F. Zhang, S. Dong, P.M. Claesson, C. Lin, J. Pan, Heating-induced enhancement of corrosion protection of carbon steel by a nanocomposite film containing mussel adhesive protein, J. Electrochem. Soc. 164 (2017) C188–C193. https://doi.org/10.1149/2.0071706jes

[19] F. Zhang, C. Chen, R. Hou, J. Li, Y. Cao, S. Dong, C. Lin, J. Pan, Investigation and application of mussel adhesive protein nanocomposite film-forming inhibitor for reinforced concrete engineering, Corros. Sci. 153 (2019) 333-340. https://doi.org/10.1016/j.corsci.2019.03.023

[20] J. Luo, X. Chen, J. Crump, H. Zhou, D.G. Davies, G. Zhou, N. Zhang, C. Jin, Interactions of fungi with concrete: Significant importance for bio-based self-healing concrete, Constr. Build. Mater. 164 (2018) 275–285. https://doi.org/10.1016/j.conbuildmat.2017.12.233

[21] N. Etteyeb, X.R. Nóvoa, Inhibition effect of some trees cultivated in arid regions against the corrosion of steel reinforcement in alkaline chloride solution, Corros. Sci. 112 (2016) 471–482. https://doi.org/10.1016/j.corsci.2016.07.016

[22] S.P. Palanisamy, G. Maheswaran, C. Kamal, G. Venkatesh, Prosopis juliflora—A green corrosion inhibitor for reinforced steel in concrete, Res. Chem. Intermed. 42 (2016) 7823–7840. https://doi.org/10.1007/s11164-016-2564-1

[23] D.B. Left, M. Zertoubi, S. Khoudali, M. Azzi, New application of chamaerops humilis l. Extract as a green corrosion inhibitor for reinforcement steel in a simulated carbonated concrete pore solution, Port. Electrochim. Acta. 36 (2018) 249–257. https://doi.org/10.4152/pea.201804249

[24] L. C, M. Rajalingam, S. George, Corrosion resistance of reinforced concrete with green corrosion, Int. J. Eng. Sci. Invent. Res. Dev. 3 (2017) 687–691.

[25] P. Shah, S. Agarwal, Aloe-Vera: A Green Corrosion Inhibitor, Www.Ijraset.Com Issue v.2 (2014) 14–17.

[26] J.O. Okeniyi, C.A. Loto, A.P.I. Popoola, Electrochemical performance of Anthocleista djalonensis on steel-reinforcement corrosion in concrete immersed in saline/marine simulating-environment, Trans. Indian Inst. Met. 67 (2014) 959–969. https://doi.org/10.1007/s12666-014-0424-5

[27] D.G. Eyu, H. Esah, C. Chukwuekezie, J. Idris, I. Mohammad, Effect of green inhibitor on the corrosion behaviour of reinforced carbon steel in concrete, ARPN J. Eng. Appl. Sci. 8 (2013) 326–332.

[28] R. Anitha, S. Chitra, V. Hemapriya, I.-M. Chung, S.-H. Kim, M. Prabakaran, Implications of eco-addition inhibitor to mitigate corrosion in reinforced steel embedded in concrete, Constr. Build. Mater. 213 (2019) 246–256. https://doi.org/10.1016/j.conbuildmat.2019.04.046

[29] S.A. Asipita, M. Ismail, M.Z.A. Majid, Z.A. Majid, C. Abdullah, J. Mirza, Green Bambusa Arundinacea leaves extract as a sustainable corrosion inhibitor in steel reinforced concrete, J. Clean. Prod. 67 (2014) 139–146. https://doi.org/10.1016/j.jclepro.2013.12.033

[30] J.O. Okeniyi, A.P.I. Popoola, E.T. Okeniyi, Cymbopogon citratus and $NaNO_2$ Behaviours in 3.5% nacl-immersed steel-reinforced concrete: implications for eco-friendly corrosion inhibitor applications for steel in concrete, Int. J. Corros. Article ID 5949042 (2018). https://doi.org/10.1155/2018/5949042

[31] M.A. Asaad, M. Ismail, M.M. Tahir, G.F. Huseien, P.B. Raja, Y.P. Asmara, Enhanced corrosion resistance of reinforced concrete: Role of emerging eco-friendly Elaeis guineensis/silver nanoparticles inhibitor, Constr. Build. Mater. 188 (2018) 555–568. https://doi.org/10.1016/j.conbuildmat.2018.08.140

[32] J.O. Okeniyi, C.A. Loto, A.P.I. Popopla, Corrosion inhibition performance of rhizophora mangle L bark-extract on concrete steel-reinforcement in industrial/microbial simulating-environment, Int. J. Electrochem. Sci. 9 (2014)

4205–4216.

[33] S.A. Abdulsada, A.I. Al-mosawi, A.A.A. Hadi, Studying the Effect of Eco-addition inhibitors on corrosion resistance of reinforced concrete, Bioprocess Eng. 1 (2017) 81.

[34] Y. Liu, Z. Song, W. Wang, L. Jiang, Y. Zhang, M. Guo, F. Song, N. Xu, Effect of ginger extract as green inhibitor on chloride-induced corrosion of carbon steel in simulated concrete pore solutions, J. Clean. Prod. 214 (2019) 298–307. https://doi.org/10.1016/j.jclepro.2018.12.299

[35] E.F. Hernández, P.F.D.J. Cano-Barrita, F.M. León-Martínez, A.A. Torres-Acosta, Performance of cactus mucilage and brown seaweed extract as a steel corrosion inhibitor in chloride contaminated alkaline media, Anti-Corros. Methods Mater. 64 (2017) 529–539. https://doi.org/10.1108/ACMM-02-2016-1646

Materials Research Forum LLC
https://doi.org/10.21741/9781644901052-8

Chapter 8

Green Corrosion Inhibitor for Electronics

S.K. Ujjain[1], P. Ahuja[1], R. Kanojia[2]*

[1]Research Initiative for Supra-Materials, Shinshu University, 4-17-1 Wakasato, Nagano-City 380-8553, Japan

[2]Department of Chemistry, Shivaji College, University of Delhi, India

*drrajnikanojia@gmail.com

Abstract

Green corrosion inhibitors used for protection of metals and alloys in electronic devices are a key interest area for researcher because of increased environmental awareness which restrict the use of toxic and hazardous corrosion inhibitors which contaminate our ecological system. Advancement of green chemical technologies towards novel vapor-phase corrosion inhibitors (VPCI) as green inhibitors for electronics corrosion and their adsorption mechanism is discussed in detail here. Also, the protective role of VPCI for various metals and alloys used in electronics components and industrial applications are discussed.

Keywords

Green Corrosion Inhibitor, Vapor-Phase Corrosion Inhibitors, Electronics, Adsorption, Protective Film

Contents

1.   Introduction

Our everyday life largely depends upon electronic equipment and devices, which is increasing day by day. From the last few decades, the electronic industry has developed as largest industry worldwide. Still, the degradation of electronic devices takes place due to many reasons, for example, electric, mechanical, magnetic and chemical effects. It may be due to simultaneous action of more than one of these effects. The main cause of chemical degradation of electronics is atmospheric corrosion which is most important among all these effects. A broad range of consequences can result because of corrosion in the perilous components of electrical automatic devices including electrical faults or breakdown of the system. Electrical components corrosion in electronic devices is more critical than corrosion of oil and gas pipelines, bridges, refineries and chemical industries because minute amount of degradation causes or affect the electrical properties (such as conductance and resistance) and hence, performance of whole electronic system. In addition, moisture and humidity are one of the initiators of undesirable effect occurred in electronics causing electrical shorts and variation in electrical resistance.

Literature survey reveals that electronic device component corrosion is a serious problem owing to which the world sustained expensive damages. The estimation is about 5 billion USD in the United Nation alone for equipment replacements and repairs [1,2]. In modern era, the electronic equipment grows in capacity but continuously shrink in size, thus, increasing the need of corrosion control. For instance, very little loss of about 1 pico gram of the material during electronic component corrosion may cause failure of device [3]. Moreover, failure of electronic devices occurs due to corrosion in different environment as reported in literature which include chemical industries, telecommunication facilities, motor vehicles, industrial equipment, medical devices, aircraft, military weapons, defense industry, automotive etc. Nowadays, our daily life is totally dependent on electronics like computers, tablet, medical equipment, car, air traffic control and also in defense missile system etc. Hence, it is essential to know some important points including: how to control electronic corrosion, which part of device is more prone to corrosion and how to maintain them.

Theory and Applications of Green Corrosion Inhibitors   Materials Research Forum LLC
Materials Research Foundations **86** (2021) 204-232   https://doi.org/10.21741/9781644901052-8

Certainly, the main issue is trying to sum up electronics corrosion because electronic components are made from various materials such as aluminum, copper, tin, nickel, silver, gold and some specific alloys etc., which interact differently in different environments. Also, these materials are affected by different types of corrosion. Gold is non-reactive and hence not susceptible to corrosion, making it useful for plating to protect other metals. For example, in switches where contact area is important and small amount of corrosion can cause faults quickly. In addition, aluminum form thick oxide film in normal environment and the resulted passive layers of oxide protects the elemental metal inside like gold plating. However, these natural and artificial passivation have their failure points. When two different types of metal come in contact with each other in an electrolyte, they form a complete circuit where ion diffuse and react with metals leading to formation of oxides and metal salts, i.e. corrosion. This process is known as galvanic corrosion. Some other types of corrosion can also occur, for example, pitting corrosion in aluminum, corrosion fatigue cracking in copper, filiform, fretting and crevice corrosion etc. [4].

The corrosion in microelectronic facilitate slow deterioration of the equipment which possess considerable threat ragging from electrical faults to complete functional breakdown that affect human being as well as economy worldwide.

2.   Causes and factors for corrosion in electronics

The critical issues in the consistency of electronics which are associated with corrosion related problems in such environment with aerial pollutants at extremely high concentrations that are compare with health associated issues. For example, the International society of automation (ISA) has recommended 10 ppbv concentration of sulfur dioxide (SO_2) in the corrosivity category as the highest limit for electronics, while 50 ppbv has been recommended by the European Environment Agency. Environments also characterized as crucial factor to cause substantial degradation of electronic components and equipment. Moisture, air composition and temperature are the main area to focus on to combat corrosion of electronics. Galvanic corrosion initiates in the presence of electrolyte, meaning some kind of moisture either in the form of drop or spray of water or in the form of relative humidity. Relative humidity is hard to control especially during storage of electronic equipment and transportation via shipping where conditions inside a container can change drastically relative to weather outside. These situations are more threatening and electronic equipment are more susceptible to corrosion than in ordinary condition. The second important factor is air composition; there are certain gases and airborne pollutant which can increases corrosion exponentially in electronics. Most common sulfur and nitrogen containing compounds which react quickly with materials of electronic components. It is considered as the prime reason of corrosion bearing in highly polluted industrial centers located in certain parts of china. Also, chloride ions are key contributor to corrosion of electronic, specifically the pitting corrosion of aluminum. The third major factor is temperature; the rate of corrosion is accelerating as temperature increased. Definitely, there is no cheapest and easiest way to control the temperature of container during storage and shipping where changing weather with time can be important factors.

2.1 Contaminant gases affect the manufacturing areas

S-containing H_2S acts as strong oxidizing agent for corrosion of silver contacts

SO_4^{2-} ions for Al/Cu

SO_4^{2-} and NO_3^- with NH_4^+ neutralization leads to crystals formation

2.2 Other problems faced in manufacturing process

Outside untreated air

Acid vapors emitted during manufacturing processes which are toxic and corrosive

2.3 Effects of ammonia

Defects produced during chemical amplified resist process

Defects produced during pattern generation processes (T-Topping)

2.4 Effects of ozone, boron and other volatile organic compounds

Defects in LCD-low temperature polysilicon

Defects in the Large-Scale Integration (LSI) and diffusion properties

Abnormal circuit voltage resistance and current leakages

Theory and Applications of Green Corrosion Inhibitors   Materials Research Forum LLC
Materials Research Foundations **86** (2021) 204-232   https://doi.org/10.21741/9781644901052-8

Table 1. The main cause and the factor that influence corrosion in electronic device and Metal component corrosion in the electronic equipment occurs at different stages.

The Main Cause of Electronic Device Corrosion		The Factor that Influence Corrosion in Electronic Device	
Reactive Gaseous Agent	NO_x, SO_x, O_3, H_2O_2, NH_3, H_2S, Volatile Organic Compounds	Meteorological Parameters	Humidity, Moisture, Temperature
Inorganic Acid	H_2SO_4, HCl, HNO_3	Corrosive Pollutants	Chloride, Sulphates, Borates, Fluorides, Sulfides
Organic Acid	Formic Acid (*Most corrosive*), Acetic Acid, Propionic Acid	Airborne Pollutant/ Corrosive	Sulfur Dioxide, Nitrogen Oxides, Hydrogen Sulfide, H_2O_2
Hygroscopic Ionic Particle	NH_4HSO_4, NH_3NO_4, $(NH_4)SO_4$	Other Compounds	Nitrogen Compounds, Organic Solvent Vapors, Resin
Environment	Tropical, Subtropical, Marine and polluted deserts etc. (electronics have high failure rates)		
Metal component corrosion in the electronic equipment occurs at different stages			
Occurrence of Corrosion in Electronic Devices During	Different Stages of Manufacturing		
	Assembly		
	Ship Transportation		
	Storage of Components and Assemblies		
	Installation and Service of Electrical Equipment		
	Field Operations of the Equipment		

Theory and Applications of Green Corrosion Inhibitors
Materials Research Foundations **86** (2021) 204-232

Materials Research Forum LLC
https://doi.org/10.21741/9781644901052-8

2.5   Airborne contamination in various sector

Airborne molecular contamination of corrosive gases inside server rooms, data centers and control rooms interrupt the functioning of electronic equipment leads to sudden failures. Contaminants percolate into data processing facilities and corrode electronic equipment leading to malfunctioning and breakdown of ACs, computers, telecom system and control panels etc. Gaseous contaminants and air pollutants such as bromine, SO_2, H_2S, chlorine, oxides of nitrogen etc. penetrate into the conditioned space of modern luxurious offices. These gases in combination with humidity or with temperature react with materials such as silver and copper in electronic circuit boards, cooling equipment, server cards, datacom connections and switch gear resulting in micro-electronic malfunction due to corrosion which leads to downtime losses. Downtime interprets as data loss, revenue loss and at last customer dissatisfaction. To name few affected by malfunctioning of electronic due to corrosion are following:

➢ Telecom Industry

➢ Distributed Control System (DCS)

➢ Data Centers

➢ Server Rooms/ IT Rooms

➢ Semiconductor

➢ Paper and Pulp

➢ Petroleum and oil Refineries

➢ Sugar

➢ Textile

2.5.1 Telecom industry

Nearby industrial areas are loaded with molecules of corrosive gases like NH_3, H_2S, SO_2, HF, Cl, O_3 etc. which are generated during the manufacturing processes and ultimately mix with the surrounding air. These airborne corrosive gases can damage the circuit boards of electronic devices that reducing the life expectancy of these sophisticated devices apart from causing malfunctions of electronics, signal failure and downtime losses.

The base transceiver station situated near industrial areas or places having heavy traffic congestion is exposed to the high risk of corrosion of their sophisticated electronic components.

2.5.2 Distributed control system (DCS)

This system is used for regulatory production processes such as petrochemicals, food & beverage, pharmaceuticals, oil refining, cement production, steel, central station power generation, paper making etc. Here, micro-corrosion can be caused by presence of corrosive gases with slight humidity. DCS servers and control rooms are equipped with sophisticated equipment like BPOS, ITEs etc., which are more susceptible to corrosive environment. If corrosion of electronic left unprocessed for long time is highly accountable for malfunctioning of electronic, retarded performance and under extreme conditions causing failure of electronic equipment.

2.5.3 Data centers

Contamination by gaseous pollutants is one of the main causes for the high rate of hardware failures in data centers. This corrosion related problem faced by computing equipment manufactures like IBM, Intel, HP, CISCO, Dell etc. Data centers located near industrial townships are more prone to gaseous contaminants which are main cause of failure of electronic equipment in data centers. Complex assembly of electronic component is made from metallic and non-metallic materials which are highly vulnerable to corrosion.

Corrosion of electronic components can affect electronic as follow:

- ➢ *Disk Drive*: Formation of disk drive in such manner that makes it highly susceptible to electronic corrosion in indoor environment.

- ➢ *Film Disk*: Information stored in disk gets lost at the reaction site. Moreover, reaction product accumulations lead to mechanical failure of data track.

- ➢ *Edge connectors of circuit board*: The connectors are made exclusively of copper or gold/nickel plated copper substrate. Corrosion initiated in the presence of moisture with gases contaminants such as Cl, SO_2, and H_2S etc. in copper substrate can lead to disruption in the connector points which affect data transfer.

3. Metals or specific alloys component for electronics

Various metals are required for designing of electronic equipment due to their electrical and physical features. Metal and alloys used in electronics are mentioned in Table 2.

Table 2. Metal and alloys used in electronics.

S. No.	Metal and Alloys	Used in Electronics
1	Gold (Au)	Gold foil or gold coating in electrical connectors, miniature circuits, hybrid, printed circuits etc.
2	Silver (Ag)	Silver used as protective coating in EMI gaskets, cables, contact relays etc.
3	Copper (Cu), Copper Alloys	Mostly used for printed circuits, cables, nuts and bolts, tablets, Radio Frequency (RF) packaging etc.
4	Aluminum (Al)	Mainly use in mounting frames, trusses, chassis, brackets, armor equipments etc.
5	Nickel (Ni)	Nickel used as coating for barrier between electric contacts of gold and copper, compatibility of dissimilar material joints, electromagnetic interference applications etc.
6	Tin (Sn)	Tin coating used between dissimilar metals for compatibility, automatic switching mechanisms, welding etc.
7	Steel, Iron (Fe)	Used for magnetic coatings and shielding for memory disks.
8	Magnesium (Mg) Alloys	Used in chassis, brackets, light structures, radar antenna dishes etc.
9	Cadmium (Cd)	Cadmium coating used as sacrificial protecting coating on iron and in electrical connectors.

Furthermore, galvanic/bimetallic corrosion ascends in metals in close contact in the presence of moisture. Following must be taken into consideration while using similar metals:

➢ Effective area of the metal coated with a companionable metal.

➢ A hermetically sealed arrangement must be used for electronic device.

➤ Sealing interfaces should be used for preventing ingress of moisture.

➤ Metal compatible packaging must be interposed between dissimilar metals.

➤ The contact areas of metal designed in such a way that the area of active anodic part should be larger than that of noble cathodic component. This can be achieved by coating to the cathode metal component.

4.   Electronic component susceptibility towards corrosion and failure analysis

4.1   Printed circuit board

Printed circuit board consists of metallic parts like the conductive patterns which are protected from humidity, corrosive gases and contaminants etc. by a tin-lead plating, passivation treatment, solder resist, an organic coating or by applying conformal coating. The microelectronic circuit boards have electrical connectors whose thickness is very much less; implies that the occurrence of smallest amount of corrosion due to humidity, pollutants and temperature etc. can cause board failure. The electrical failure is attributed to corrosion by pollutants such as sulfur compounds. The electronic circuit board is adequately affected by atmospheric corrosion which are caused by sulfur species. Most of the times awfully tarnished circuit board pads and connectors can be found visually or by microscopic examination. Corrosion products of heavy copper sulfide (Cu_2S) are present on many conductor tracks and pads, while gold-plated external connector pins in the circuit boards are not corroded.

The open connector pads of the electronic module show the growth of iridescent black deposits which is analyzing by optical microscopy. The brown black deposits appeared due to the corrosion of copper in sulfur-bearing environment. These corrosion deposits growing out with the passage of time beneath connector holes which leads to a short circuit of electronic components. Also, similar kind of corrosion failure of electronic equipment take place in sealed field units of industrial plant. This happens because of corrosive atmosphere which migrated inside the sealed units that causes multiple repetitive failures of electronic equipment. This arises in electronic circuit board due to corrosion fatigue of copper conductor wires. In addition, corrosion fatigue is the main reason of failure of the transformer copper conductor leads. In view of the fact that, sulfur-bearing environment is a promoter of the fatigue cracking. The leads fracture surface in failed circuit boards showed low ductility, transgranular fatigue cracking with tarnished or blackened tiny cracks contained dark colored corrosion products.

Silver and copper are extensively utilized in electronics owing to their excellent electrical and thermal properties. But, both show very low activation energies favorable for the formation of sulfides with hydrogen sulfide. The sulfide containing corrosion products are permeable in nature that eases hydrogen sulfide to reach the surface of metal. The gas or humidity is present in environment which accelerates the corrosion process; there is no mechanism to stop it. In contrast, another commonly used metal in electronic assembly is tin; an impermeable sulfide layer is formed, which stops further reaction. Moreover, the transformer copper wires are tinned that provides some initial protection to electronic. The fatigue cracks start growing in the thin tin layer which exposure bare copper in wires to the sulfur containing corrosive environment [5-9].

4.2 Contact and connector

Contact and connector consists of mainly brass, copper or different alloys plated with nickel or gold interlayer. They are exposed to environmental pollutant or corrosive gases deprived of any shield and thus, are extremely vulnerable to corrosion.

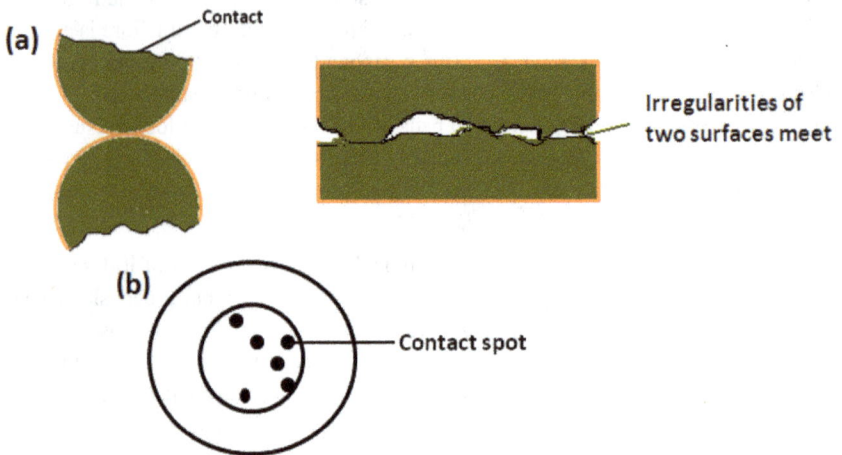

Figure 1. (a) Shows contact surface irregularities and (b) current flow through contact spots.

Theory and Applications of Green Corrosion Inhibitors   Materials Research Forum LLC
Materials Research Foundations **86** (2021) 204-232   https://doi.org/10.21741/9781644901052-8

The two surfaces are in contact with each other; meet at a point where the surfaces show some irregularities and therefore, proper electrical contact happens at minor spots as shown in Figure 1 (a & b). The current crosses through such minor contact spots causes a voltage drop. As a result of contact surface corrosion, resultant products formed a uniform film leading to supplementary voltage drop. Further, other resultant products formed reduces the contact spots, increasing the danger of recurrent contact failures as shown in Fig. 2.

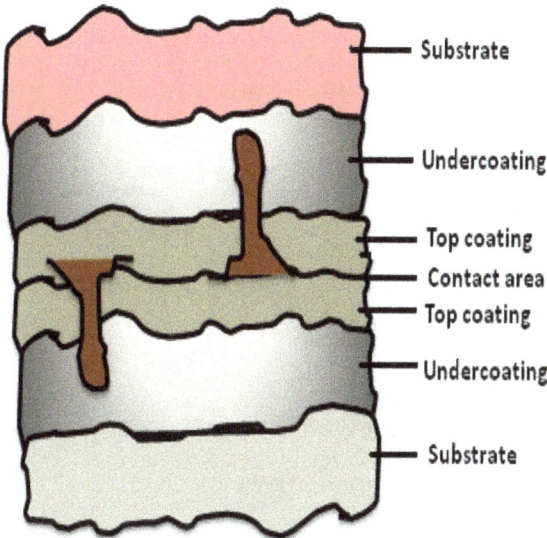

Figure 2. Schematic representation of corrosion products and their effect on contact area.

4.2.1 Pore corrosion in electrical contacts

Pore corrosion mainly witnessed in gold coated electrical contacts is a form of bimetallic corrosion. The top coating of noble metal coating exhibits no pores, but additional defect in the protective coating that reach beneath the coating and subsequently to the substrate. It creates a galvanic set up in which corrosion of electrode (copper or silver) occurs and resulted decomposition products block the pores, drift towards the pore walls with finally destined to the surface, resulting in increased contact resistance as demonstrated in Fig. 3.

Since, pores in the top coating which makes fairly short life of gold coated contact in severe environments.

In addition, pore corrosion occurs on surfaces having gold plating mainly due to nickel/copper/copper alloy undercoating corrosion, which is stimulated in the presence of humidity, H_2S, Cl_2 and SO_2. Resulted decomposition products from corrosion formed at edges of the effective surface passes through pores to the noble metal and develop a shielding film over the effective contact region. Interestingly, products resulted from corrosion of copper creep on both silver and gold coating, while silver corrosion product creeps only over the gold coated region. Nickel mainly used as an undercoating to reduce the copper corrosion products creep. However, the corrosion products creep is higher on nickel compared to copper and is stimulated by the presence of H_2S.

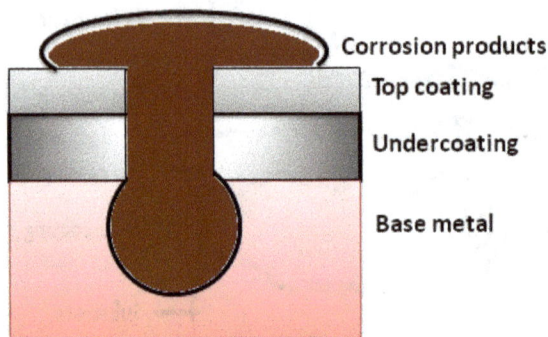

Figure 3. A schematic illustration of pore corrosion.

4.2.2   Fretting corrosion of electronic connectors

Fretting corrosion occurs due to low amplitude oscillatory motion of two surfaces in stationary contacts or oscillatory slip among binary vibrating surfaces in contact. Such relative motion can happen due to mechanical/thermal movements or else by repetitive electrical and magnetic vibrations. As a consequence of fretting corrosion, an insulating film is formed that detach the surfaces electrically to increase their contact resistance. Material like tin can easily form oxides and are highly prone towards fretting corrosion.

In short, it can be said that failure of contacts and connectors is largely related to the material of choice and environmental factors. The gold-plated surface is mostly susceptible towards pore corrosion and its products creep, whereas fretting corrosion is the dominant failure mechanisms for tin plated surface [10,11].

4.3   Integrated circuits

Integrated circuit consists of an extensive metallic materials and narrowly spaced conductive patterns with essential resistors, chips and additional components. The materials used as conductive pattern are copper, aluminum or aluminum alloys which are coated with gold, silver, palladium or tin. These metallic conductive paths are very prone to electrolytic dissolution, ionic migration, pitting corrosion or galvanic corrosion in presence of humidity and environmental pollutants [12]. Aluminum conductors show pitting corrosion and electrolytic dissolution because of difference in the electrical potential amongst the separated conductive paths. Due to the presence of moisture, the conductor which show higher electrical potential act as anode and its dissolution may occur [13]. When contaminants like chloride are present, the dissolution of aluminum proceeds as similar manner like pitting corrosion and failure of electronic devices may occur because of gap in the aluminum conductive path. Further, the copper conductors can also dissolve as anode, resulting in devices failure due to electrolytic dissolution and pitting corrosion. Sometimes, ionic migration may happen due to minor potential difference amongst the conductors and the distance between the paths is too small which generates high electric field strengths. In addition, dendritic pattern can be formed due to metal ions migration towards polarized cathodic conductor [14]. Such dendritic pattern is large as compared to dissolved metal volume, which is main cause of a short circuit amongst the adjacent conductive paths. Metals include tin, copper, gold, lead and silver are more susceptible to electrochemical migration.

Furthermore, a galvanic couple is formed when electric contact is established between two dissimilar metals in the humid environment, which is termed as bimetallic corrosion or galvanic corrosion. An aluminum bond pads coupled to gold wire is common form of integrated circuit galvanic corrosion [15].

4.4   Solder corrosion: the corrosive effect of soldering flux

The soldering operation successfully done will depend upon several factors. Suitable choice of soldering flux is very important because the flux medium is used to provide initially transfer of heat from the hot-soldering iron to the surfaces being join together. During the assembly process, there is no requirement of highly activated flux because according to soldering specification limit, the interconnect material exhibits excellent solderability. But, according to corrosion point of view, the flux of high activity is more harmful and used during the initial pre-tinning of component forms nickel alloy to accomplish solderability. Afterward pre-tinning, the flux deposits cleaned from the component preclude corrosive attack as a function of time

The failure of several gold-plated lead over flat packaged part is serious corrosion related problem. De-golding and pre-tanning are achieved by dipping the leads in solder baths flux having mildly activated rosin. After stored for months, it can be observed that the leads show fracture completely when exposed to light handling operations. The main reason of failure is stress corrosion which initiates a crack in the lead material and spread through a fatigue mechanism [16].

4.5   Hermetic packages

Hermetic package provides an extra protection to electronic devices from environmental impacts. Hermetic seals are delicate and a small crack occurs due to handling or by joining circuit boards which create leaks. Moisture adsorb on the wall of the package which is the first step of initiation of corrosion of parts or components or equipment after that the failure of electronic devices occur [17].

The biggest challenge faced by electronics manufacturing industries is effective corrosion protection by packaging. But, electronics items/devices may experience corrosion failure because of extreme temperature variations or moisture condensation during transit or moisture trapped on the inner wall of the packaging that builds up during shipment. For instance, a telecommunication company suffered packaging failure of 86 % while shipping to the far east from north America. Nguyen and coworker investigated moisture permeation in plastic packages during capacitance monitoring corroded the equipment. The repair and return of such corroded equipment increased the loss of the company [18].

5.   Reliability and cleanliness

Electronic manufacturers take numerous precautions to keep the system, equipment, components and premises contaminant-free and clean to safeguard the complete removal of microscopic contaminants from components/equipment of electronic devices which can foster an initiation corrosion site resulting in electronic devices failure. However, discoloration of parts or components of electronic and corrosion can still occur in the highly controlled area of a plant.

6.   Electronics corrosion protection

When analyzing any method to control corrosion in industrial system or operational processes should comprise such things as the durability, coating scheduled maintenance, service life, sustainability and life cycle costs. The selection of the most effective corrosion protection system is important. The corrosion protection systems are galvanizing, painting, coating or use of inhibitors as organic, inorganic, adsorption–type, anodic, cathodic, mixed inhibitor, environmental friendly inhibitor and green volatile or vapor phase corrosion inhibitor. Most compounds are toxic and cause environmental hazards. Due to hazardous effects of toxic compounds, researchers are more focused on the use of non-toxic compounds.

An important aspect of manufacturing electronics industry is protection of equipment from corrosion and to ensure that components and equipment are clean and corrosion free. In addition, taking all preventive care, corrosion is biggest problem during manufacturing as well as post shipment. The vapor phase corrosion inhibitor (VPCI) technology has been introduced as most reliable technology for effective corrosion prevention.

7. Vapor phase corrosion inhibitor (VPCI) technology

VPCI is an innovative technique for opposing corrosion of electronics. Mostly, amine salts of carboxylic acids are used as vapor phase corrosion inhibitor which deposits as microscopic protective film on the metal/alloy substrate of parts or components of electronic those exhibit molecular attraction and are self-regulated by equilibrium towards the surface of metal/alloy. The VPCI molecules are so minute and hence do not obstruct the functioning of electronics. Molecules of VPCI are deposited using source material vaporization and dispersion all over in closed space till equilibration is attained. As VPCI molecules are in contact with metal surface, it adsorbs onto surface of metal and form highly effective protective-film/layer which prevent corrosion of electronic devices (Fig. 4). If the closed space is slightly opened and letting few molecules of VPCI to elope, simultaneously, more molecule from the source material can replace any escaped molecules to regenerate a new shielding film onto the metal surface. The most important feature in adapting VPCI method is its electrostatic discharge (ESD) protective capabilities to shield electronics. In absence of ESD, electronics can be easily corroded owing to friction while shipment.

Furthermore, VPCI are low cost and effective way to protect enclosed spaces such as electrical cabinets and can also be sprayed over the circuit boards and electrical contacts to generate a shielding layer over them without hampering their electrical conductivity. In addition, VPCI plays pivotal role as green corrosion inhibitors and are essentially required while packaging electronics, coatings, reinforced concrete and industrial processing because of their safe handling and nontoxic phosphates & halogens free properties. Nano-sized VPCI derived from agriculture by-products maintains ecological

Theory and Applications of Green Corrosion Inhibitors | Materials Research Forum LLC
Materials Research Foundations **86** (2021) 204-232 | https://doi.org/10.21741/9781644901052-8

balance and its effectiveness is related to the diffusion of molecules, active areas and void spaces filling by protective vapor molecules.

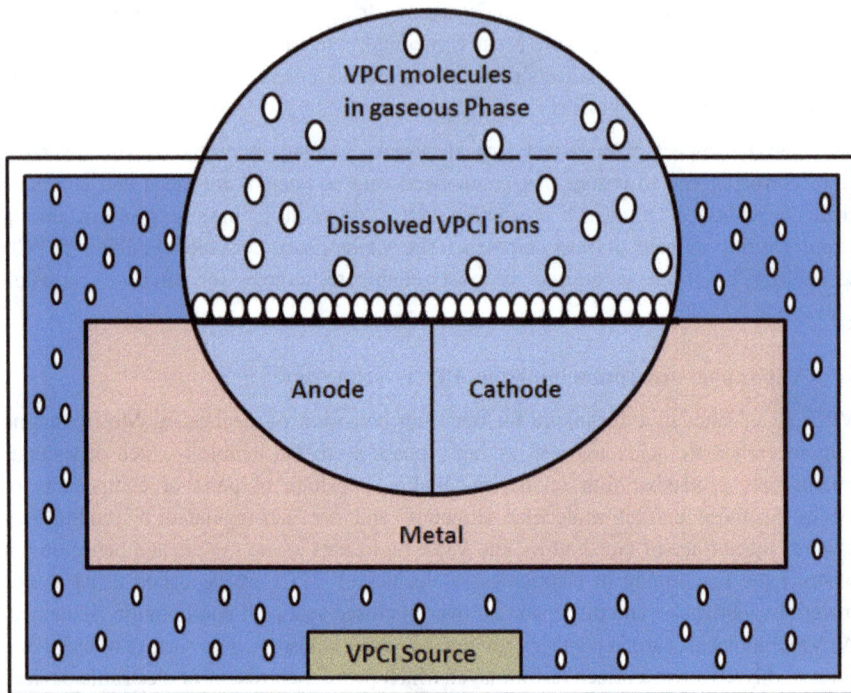

Figure 4. Schematic illustration of vaporization and dispersion of VPCI molecules from their source material throughout the enclosed space. VPCI molecules get adsorb onto the metal surface to form highly effective protective film.

VPCI are considered as a class of inhibitor which can be utilized to protect the corrosion/oxidation of metals or alloys where other surface treatments are impractical. Numerous VPCI are reported as effectual inhibitors for copper, aluminum, zinc, iron and their alloys. Weak and volatile acids or bases in their salt form (having vapor pressure 10^{-2}-10^{-7} mmHg) provides a source material for VPCI for different metals. Literature studies revealed that VPCI protect corrosion of metals or alloys either forming a protective layer onto the surface of metal or by neutralizing the corrosive reagents like moisture, H_2S, SO_2 and CO_2 etc. The VPCI can be utilized in form of sprays, powders, coatings, adhesives, foams etc. They can be easily vaporized and condensed via adsorb onto the space vacant in the crevices, pores and cracks of metallic surfaces and thus gives complete protection. Few VPCI are also reported earlier whose hydrolyzed products can be considered as efficient inhibitors of corrosion [19-21]. For example, nitrile–carbamide and nitrite–hexamethylenetetramine act effectively for metallic corrosion due to the establishment of extremely unstable ammonium nitrite in the course of hydrolysis reaction [22, 23].

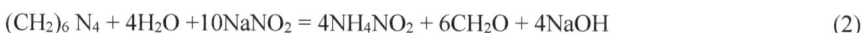

$$CO(NH_2)_2 + 2H_2O + 2NaNO_2 = Na_2CO_3 + 2NH_4NO_2 \tag{1}$$

$$(CH_2)_6\,N_4 + 4H_2O + 10NaNO_2 = 4NH_4NO_2 + 6CH_2O + 4NaOH \tag{2}$$

In general, vaporization of the VPCI take place without dissociating their molecules, however, cyclohexylamine and carbon dioxide vaporization happens mostly via hydrolytic dissociation or thermal process [24].

8.   Vapor pressure measurement by various methods

The adsorption of VPCI molecules onto metallic surface is more favorable in the presence of moisture or humidity due to increased volatility (saturation vapor pressure represented as P^0) of the components through hydrolysis. However, P^0 decreases with increase in humidity are reported in various cases. In addition, humidity and temperature are important parameter limiting the volatility of the VPCI. There are various methods for the measurement of P^0 and can be classified as dynamic, boiling point, open surface evaporation, static measurement and improved forms.

8.1   Regnault dynamic method

This method implicates heating a fixed amount of inhibitor under flowing stream of oxygen free nitrogen at the constant temperature and subsequent weight loss is measured. The rate of vaporization of the VPCI molecule at various temperature can be calculated by using the equation shown below [25]:

$$\frac{Total\ volume\ of\ compound}{Volume\ of\ vapor} = \frac{Total\ pressure}{Vapor\ pressure} \tag{3}$$

The saturation vapor pressure is derived from the plot of temperature and vapor pressure.

Theory and Applications of Green Corrosion Inhibitors   Materials Research Forum LLC
Materials Research Foundations **86** (2021) 204-232   https://doi.org/10.21741/9781644901052-8

8.2   Boiling point determination method

This method employs to determine compound's boiling temperature or pressure, provided saturation vapor pressure (P^0) becomes equal to external pressure (P).

8.3   Knudsen effusion method

This method used for the evaluating the gas effusion rate via small orifice. The vapor pressure can be evaluated by following equation [24, 25]:

$$P_t^{\ 0} = \frac{\Delta W}{KSt} \left(\frac{T}{M}\right)^{\frac{1}{2}} \tag{4}$$

Where, ΔW denotes compound weight loss, K constant, S represents orifice area at time t and absolute temperature T, M is compound molecular weight.

8.4   Microbalance method

This method involves the weight loss quantification via extremely precise microbalance calibrated with compound of known vapor pressure at a particular temperature.

8.5   Torsion effusion method

This method used for determination of recoil force developed when gaseous form of VPCI molecules effuse from pores to vacuum. The source compound is placed in an effusive container and suspended by fiber. The resulting vapor escaped through a path with two oddly located pores leading the twisted fiber with particular angle. The vapor pressure degree can be calculated from the deflection angle by means of equation [24, 25]:

$$P_t^{\ 0} = \frac{2KB}{f_1 d_1 a_1 + f_2 d_2 a_2} \tag{5}$$

Where, P^0 denotes the vapor pressure, B is the torque steady angle generated at temperature (T) due to effusion, d_1 and d_2 symbolize the distance between two pores from the axis of origin, f_1 and f_2 denote the correction factors and a_1 and a_2 represent the area of the cavities.

Due to difficulty in determination of adequately precise d, a, f and K factors, an improved equation (6) has been suggested for which the measurement instrument is calibrated at various temperatures using a compound of known vapor pressure

$$P_t = K_t B \qquad (6)$$

Where, K_t is constant utilized to evaluate other compounds vapor pressure under the identical range of temperature.

The correlation between electronic and physicochemical features of several metals with inhibitor's molecular parameters envisages performance and inhibition efficiency of VPCI.

9.   Effect of temperature on the vapor pressure

The saturation vapor pressure (P^0) of VPCI molecules depends on molecular structure as well as their molecular weight. Literature studies reveal that the volatility of the organic compound decreases with increase in the molecular mass. Further, evaporation mechanism and composition showed dependency on intrinsic characteristics of the VPCI alongwith some external parameters. Among them, the temperature has a significant role on the VPCI vapor pressure. The Claussius–Clapeyron equation is used to analyzing the effect of temperature [26].

$$\log P = A - \frac{B}{T} \qquad (7)$$

Where, P is the vapor pressure of compound, A and B are constants at absolute temperature T.

The P^0 decreases with increase in latent heat of the compound owing to enhanced intermolecular force of attraction, leading to augmented force of attraction (Van der Waals force).

10.   Effect of pH

The VPCI molecules metallic corrosion inhibition mechanism is a two-step process involving VPCI molecules vaporization and their corresponding dissolution in existing moisture film. This consequently protects the occurrence of corrosion.

VPCI molecules have propensity to improve hydrophobic nature of the moisture film over metal or alloy surface, which plays pivotal role in corrosion inhibition. Dissolution of the molecules of VPCI occurs when the pH of the moisture is not very low (acidic) or very high (basic). The study reveals weak volatile acids and bases salts as appropriate VPCI. Most of the VPCI demonstrate their superior activities between 5.5 to 8.5 pH range [27, 28]. The protective film forming nature of VPCI molecule can protect the metallic surface from corrosion and also defusing the environmental corrosive agents like SO_2, H_2O and CO_2 etc.

11. Types of vapor phase corrosion inhibitors (VPCI)

VPCI are impeding corrosion in closed spaces via vapors condensation in microscopic crystals form, which dissolves the moisture layer present on the metal surface. The dissolved VPCI ions displaces water molecules from the surface of metal and forming monomolecular protection layer which retarding the corrosion rate. VPCI may be added to a variety of package materials: polymer film (e.g. low-density polyethylene), powder, paper, foam, oils etc. The following organic compounds are used as vapor phase corrosion inhibitors (VPCI):

- ➢ Cyclohexylamine
- ➢ Dicyclohexylamine
- ➢ Guanidine
- ➢ Amino alcohols
- ➢ Nitrites

Different types of organic inhibitors studied to evaluate their efficiency as VPCI on electronic substrates is given below:

- ➢ *Cyclic amine carboxylates* used to protect components of circuit board in the presence of H_2S
- ➢ *Cyclic amine carboxylated nitrate* provide protection to components of electronics devices under high humidity conditions
- ➢ *Triazoles* protects bolts and solder of electronics
- ➢ *Alcoholic amine carboxylate* enhances or accelerates the diodes corrosion
- ➢ *Cyclic amine carboxylated fatty acids* increases the corrosion of bolts and solder

Therefore, proper evaluation of corrosion inhibitors is fundamental requirement for protection of components of electronic devices. Various VPCI in different environment reported in literature are mentioned in Table 3.

Table 3. Various VPCI in different environment.

S. No.	Green Vapor Phase Inhibitor	Environment	Metals or Alloys
1	Dicyclohexylammonium nitrite [29]	High relative humidity (RH), hydrogen chloride and sulphide contaminated environment	Copper and electronic items
2	Laurichydrazide alongwith other acids say cinnamic, succinic acid, etc. [30]	100% RH	Copper, brass, zinc, aluminum & mild steel
3	1-(2-Aminoethyl)-2-dec-9-enyl-2-imidazoline salicylic acid [31]	-	Brass, copper, aluminium & mild steel
4	6-Amino-benzothiazole [32]	100% RH	Copper, mild steel
5	β-Amino alcohols [33]	100% RH	Brass
6	Benzoic hydrazide stearate (BHS) [34]	SO_2 and NaCl, high RH & high temperature.	Copper, brass & mild steel
7	1,3-bis-diethylamino-propan-2-ol (VPCI) [35]	100% RH	Brass
8	Thiomorpholin-4-methyl-phosphonic acid [36]	-	Carbon steel
9	Amine carboxylate [37]	-	Carbon steel
10	Piperazine and its derivatives [38]	100% RH at room temperature for 16 hours & 40°C for 8 hours	Mild steel
11	Triethylamine (TEA) [39]	-	Mild steel
12	n-Caprylic acid (CA) and 2-amino benzothiazole (ABT) [40]	40°C temperature	Mild steel

13	Diallylamine (DAA), and 1-benzylimidazole (1-BIZ) [41]	50°C temperature	Mild steel
14	Thiocarbohydrazides and their salts [42]	Vapor phase	Mild steel
15	Oleic hydrazide benzoate (OHB)[43]	NaCl	Mild steel
16	Dicyclohexylamine [44]	CO_2	Mild steel
17	Urea–amine [45]	High humidity	Mild steel
18	Diethylamine phosphate (VPCI) [46]		Steel components

12. Analysis of corrosion by different method

12.1 Vapor pressure determination

A liquid or solid substance is in equilibrium with its vapor at temperatures T and pressures P is known as the vapor pressure of the liquid or solid. Knudsen method is utilized for determining the vapor pressure of VPCI. For VPCI, the chosen organic compound should not have vapor pressure very high or low [47, 48].

$$P = \frac{W}{At}\left[\frac{2\pi RT}{M}\right]^{\frac{1}{2}} \tag{8}$$

Where, P is the VPCI vapor pressure (mmHg), A is the orifice area (m^2), t is the exposure time (second), W is the substance weight loss (Kg), T is the temperature (K), M is the inhibitor molecular mass (Kg) and R is the universal gas constant.

12.2 Weight loss method

Metal or alloy specimen is polished with emery papers of various grading like 100, 200, 400 and 600, followed by thorough washing with distilled water, ethanol and acetone. Specimen is dried and stored in desiccators for 1 day. After weighing initial weight of specimen, it is placed in air thermostat chamber with fixed amount of VPCI kept at 50°C for 24 hours (exposure time). Afterward, metal specimen surface which gets coated with uniform VPCI adsorbed film is moved to a digitally controlled humidity chamber kept for 10 days at 85% RH and 50°C [49]. Again, specimen is weighted and weight loss of specimen and efficiency of inhibitor is derived using following equation:

$$IE\,(\%) = \frac{w_0 - w_i}{w_0} \times 100 \tag{9}$$

Where, W_i and W_o is the weight loss value in the presence & absence of inhibitor and inhibition efficiency (IE) of the inhibitor can be determine.

12.3   Esckhe method

This method also involves use of the pre-weighed polished specimen. Kraft paper is immersed in VPCI (30 seconds), followed by drying so that even film of VPCI adsorbed over it. After that, the specimen is enveloped with VPCI soaked Kraft paper and placed in a humidity chamber (85% RH) at 50°C for first 12 hours and only at 50°C for subsequent 12 hours, following the same procedure for 10 days. This cycle initially developed vapors followed by their condensation onto the specimen surface regularly. Next, the specimen is treated in identical mean as treated in weight loss method to determine the rate of corrosion and inhibition efficiency (IE) [50].

12.4   Salt spray method

The pre-weighed metal or alloy specimen is exposed to VPCI in air thermostat for 1 day and then the specimen is placed into salt spray chamber with 3.0 wt.% NaCl solution kept at 50°C for 10 days along with non-VPCI treated specimen. After 10 days, the specimens exposed to sodium chloride spray is treated as previously discussed weight loss method to determine rate of corrosion and inhibition efficiency (IE).

The corrosion inhibition ability and adsorption properties of VPCI on the metallic surface can be analyzed by various employed techniques such as polarographic, electrocapilarity, radiotracer, potentiodynamic polarization, electrochemical impedance spectroscopy (EIS) and contact angle methods discussed in other chapters.

13.   Advantages of VPCI

The VPCI molecules are used in various sectors namely in the surgical equipment, military devices, nuclear plants, automotive industries, petrochemical plants, owing to their propensity to actively adsorb on the contact area of metal or alloy. The molecules of VPCI have high vaporization and condensation ability resulting in less corrosion susceptible metallic surfaces. The vapor phase corrosion inhibitors have various advantages as given below:

- ➤ Ease to inserted into foams, spray, powders, adhesives, coatings and plastics
- ➤ Ease of utilizing it during packaging to impede stored items corrosion
- ➤ Molecules vaporize easily and vapor covers the entire surfaces involving pores, cracks and crevices where usual inhibitors cannot infiltrate

> VPCI owing to their thin nature, do not alter the texture of surface and can be easily removed without hampering metallic surface properties

> Ease of utilizing them during the transportation of metals and electronic components

This chapter discussed VPCI as an effective green corrosion inhibitor for electronics. Also, it presented a comparative discussion on the advantages of green VPCI over conventional corrosion inhibitors towards the metallic or alloy corrosion in electronics. As a final point, attention has been given to the VPCI adsorption behavior. The inclination towards VPCI green nano-inhibitors is an innovative area for future research.

References

[1] J. D. Gutten, Corrosion in electronic industry, Corros. 13 (1987) 1107-1112.

[2] B.A. Miksic, P.J. Martin, Proceedings of 6th European Symposium on Corrosion Inhibitors, Ferrara, Italy, 1985, 941-950.

[3] J.S. Vimala, M. Natesan, S. Rajendran, Corrosion and protection of electronic components in different environmental conditions – an overview, The Open Corros. J. 2 (2009) 105-113. https://doi.org/10.2174/1876503300902010105

[4] W.H. Abbott, The Effects of Operating Environments on Electrical and Electronic Equipment Reliability in the Pulp and Paper Industry, presented at the IEEE Industry Applications Society 1983 Pulp and Paper Technical Conference, May 1983, IEEE Conference Record, New York, Institute of Electrical and Electronic Engineers, Inc, 1983.

[5] A. Al-Hashem, Atmospheric Corrosion Map of Kuwait for the Alloys of Carbon Steel, Copper and Galvanized Steel, presented at the 11th Middle East Corrosion Conference, Bahrain Society of Engineers & NACE International, 2006.

[6] P. L. Martin, Electronic Failure Analysis Handbook, Chapter 1: Overview of Electronic Component Reliability, McGraw- Hill, 1999.

[7] G.R. Lobley, R.D. Tems, Learning from Environmental Cracking Failures, Presented at the 11th Middle East Corrosion Conference, Bahrain Society of Engineers & NACE International, 2006.

[8] C.O. Muller, Multiple contaminant gas effects on electronic equipment corrosion, Corros. J. Sci. Eng. 47 (2) (1991) 146-151. https://doi.org/10.5006/1.3585230

[9] R.D. Tems, G.R. Lobley, S. Mehta, Corrosion of Electronic Control System in Gas Treating Environments, paper presented in NACE International Conference & Expo, Corros. (2007) 7401-7415

[10] R. Ambat, S.G. Jensen, P. Moller, Corrosion reliability of electronic systems, ECS Trans. 6 (2008) 17-28. https://doi.org/10.1149/1.2900650

[11] J. Henriksen, R. Hienonen, T. Imrell, C. Leygraf, L. Sjogren, Corrosion of Electronics: A Handbook Based on Experiences from a Nordic Research Project, Bulletin No. 102, Swedish Corrosion Institute, Stockholm, Sweden, 1991.

[12] G. Frankel, Corrosion of Microelectronic and Magnetic Data storage Devices in Corrosion Mechanisms in Theory and Practice, 3rd edition, ed. P. Marcus, CRC press/ Taylor & Francis Group, Boca Raton, 2012, 825-861. https://doi.org/10.1201/b11020-20

[13] R.B. Comizzoli, R.P. Frankenthal, R.E. Lobnig, G.A. Peins, L.A. Psota-Kelty, D.J. Siconolfi, J.D. Sinclair, Corrosion of electronic materials and devices by submicron atmospheric particles, Electrochem. Soc. Interface 2(3) (1993) 26-34.

[14] D. Minzari, M.S. Jellesen, P. Moller, R. Ambat, On the electrochemical migration mechanism of tin in electronics, Corros. Sci. 53 (2013) 3366-3379. https://doi.org/10.1016/j.corsci.2011.06.015

[15] O.V. Elisseeva, A. Bruhn, J. Cerezo, A. Mavinkurve, R.T.H. Rongen, G.M. O'Halloren, R. Ambat, H. Terryn, J.M. Mol, Novel electrochemical approach to study mechanism of Al-Au wire- bond pad interconnections, Corros. Eng. Sci. Tech. 48 (2013) 409-417. https://doi.org/10.1179/1743278212Y.0000000067

[16] B.D. Dunn, C. Chandler, The corrosive effect of soldering fluxes and handling on some electronic materials, Welding Research Supplement, 1980, 289-307.

[17] R.R. Tummala, E.J. Rymaszewski, Microelectronics Packaging Handbook, Van Nostrand Reinhold, New York, 1989. https://doi.org/10.1007/978-1-4613-1069-3

[18] T. Nguyen, C.A. Kovac, Moisture diffusion in electronic packages, SAMPE Electronic Materials & Processes Conferences, June 1987, 574-589

[19] A. Subramanian, M. Natesan, V.S. Muralidharan, K. Balakrishnan, T. Vasudevan, An overview: vapor phase corrosion inhibitors, Corros. 56 (2) (2000) 144-155. https://doi.org/10.5006/1.3280530

[20] E. Vuorinen, E. Kálmán, W. Focke, Introduction to vapor phase corrosion inhibitors in metal packaging, Surf. Eng. 20 (4) (2004) 281-284. https://doi.org/10.1179/026708404225016481

[21] B.A. Miksic, A. Vignetti, Vapor Corrosion Inhibitors: Successful Field Applications in Electronics, NACE Corrosion Conference, March 2000, Orlando, 2000, 7.

[22] D. Bastidas, E. Cano, E. Mora, Volatile corrosion inhibitors: A review, AntiCorros. Meth. Mater. 52 (2005) 71–77. https://doi.org/10.1108/00035590510584771

[23] N.N. Andreev, Y.I. Kuznetsov, Physicochemical aspects of the action of volatile metal corrosion inhibitors, Russ. Chem. Rev. 74 (2005) 685–695. https://doi.org/10.1070/RC2005v074n08ABEH001162

[24] Y.I. Kuznetsov, Current state of the theory of metal corrosion inhibition, Prot. Met. 38 (2002) 103–111. https://doi.org/10.1023/A:1014904830050

[25] I. Rosenfeld, B. Persiantseva, P. Terentiev, Mechanism of metal protection by volatile inhibitors, Corros. 20 (1964) 222t–234t. https://doi.org/10.5006/0010-9312-20.7.222t

[26] J. Bastidas, B. Chico, M. Alonso, E. Mora, Corrosion of bronze by acetic and formic acid vapors, sulphur dioxide and sodium chloride particles, Mater. Corros. 46 (1995) 515–519. https://doi.org/10.1002/maco.19950460902

[27] A.I. Altsybeeva, V.V. Burlov, N.S. Fedorova, T.M. Kuzinova, G.F. Palatik, Volatile inhibitors of atmospheric corrosion of ferrous and nonferrous metals. I. Physical and chemical aspects of selection of starting reagents and synthetic routes, Int. J. Corros. Scale Inhib. 1(1) (2012) 51–64. https://doi.org/10.17675/2305-6894-2012-1-1-051-064

[28] A.I. Altsybeeva, V.V. Burlov, N.S. Fedorova, T.M. Kuzinova, G.F. Palatik, Volatile inhibitors of atmospheric corrosion of ferrous and nonferrous metals. II. Prediction of the efficiency of volatile inhibitors of atmospheric corrosion of steel (with Schiff and Mannich bases as examples), Int. J. Corros. Scale Inhib. 1(1) (2012) 99–106. https://doi.org/10.17675/2305-6894-2012-1-2-099-106

[29] M.E. Tarvin, B.A. Miksic, Volatile corrosion inhibitors for protection of electronics, Conference of National Association of Corrosion Engineers (NACE), 1989, 9.

[30] M.A. Quraishi, V. Bhardwaj, J. Rawat, Prevention of metallic corrosion by lauric hydrazide and its salts under vapor phase conditions, J. Am. Oil Chem. Soc. 79 (2002) 603–609. https://doi.org/10.1007/s11746-002-0530-6

[31] M.A. Quraishi, D. Jamal, V. Bhardwaj, Development and testing of new volatile corrosion inhibitors for multimetal systems, Mater. Perform. 42 (2003) 44–48.

Materials Research Forum LLC
https://doi.org/10.21741/9781644901052-8

[32] J. Rawat, M.A. Quraishi, Influence of some 6-methoxy-aminobenzothiazole derivatives on corrosion of ferrous and nonferrous metals under vapor-phase conditions, Corros. 59 (2003) 238–241. https://doi.org/10.5006/1.3277556

[33] G. Gao, C. Liang, Some β-amino alcohols compounds as green volatile corrosion inhibitors for brass, J. Electrochem. Soc. 154 (2007) C144–C151. https://doi.org/10.1149/1.2405868

[34] M.A. Quraishi, D. Jamal, Inhibition of metals corrosion by a new vapor phase corrosion inhibitor, J. Metall. Mater. Sci. 47 (2005) 45–50.

[35] G. Gao, C.H. Liang, 1, 3-Bis-diethylamino-propan-2-ol as volatile corrosion inhibitor for brass, Corros. Sci. 49 (2007) 3479–3493. https://doi.org/10.1016/j.corsci.2007.03.030

[36] H. Amar, T. Braisaz, D. Villemin, B. Moreau, Thiomorpholin-4-ylmethylphosphonic acid and morpholin-4-methyl-phosphonic acid as corrosion inhibitors for carbon steel in natural seawater, Mater. Chem. Phys. 110 (2008) 1–6. https://doi.org/10.1016/j.matchemphys.2007.10.001

[37] E. Vuorinen, W. Skinner, Amine carboxylates as vapor phase corrosion inhibitors, Brit. Corros. J. 37 (2002) 159–160. https://doi.org/10.1179/000705902225002385

[38] M.A. Quraishi, F. Ansari, J. Rawat, Investigation of piperazine and its derivatives as vapor phase corrosion inhibitor for mild steel, Open Electrochem. J. 1 (2009) 32-36. https://doi.org/10.2174/1876505X00901010032

[39] N.N. Andreev, Y.I. Kuznetsov, Volatile inhibitors of metal corrosion. II. Interaction of systems being protected with the environment and corrosion prevention conditions, Int. J. Corros. Scale Inhib. 1 (2) (2012) 146–153. https://doi.org/10.17675/2305-6894-2012-1-2-146-153

[40] H. Kumar, V. Saini, V. Yadav, Study of vapor phase corrosion inhibitors for mild steel under different atmospheric conditions, Int. J. Eng. & Inn. Tech. 3(3) (2013) 206–211.

[41] A.I. Altsybeeva, V.V. Burlov, N.S. Fedorova, T.M. Kuzinova, Volatile inhibitors of atmospheric corrosion of ferrous and nonferrous metals. IV. Application of the VNKh-L-408 inhibitor in an electrostatic field, Int. J. Corros. Scale Inhib. 2 (3) (2013) 194–202. https://doi.org/10.17675/2305-6894-2013-2-3-194-202

[42] F. Ansari, M.A. Quraishi, Prevention of metallic corrosion by thiocarbohydrazides and their salts in vapor phase environment, Chem. Eng. Commun. 198 (2010) 61–72. https://doi.org/10.1080/00986445.2010.493120

[43] Sudheer, M.A. Quraishi, E.E. Ebenso, M. Natesan, Inhibition of atmospheric corrosion of mild steel by new green inhibitors under vapor phase condition, Int. J. Electrochem. Sci. 7 (2012) 7463–7475.

[44] D.Q. Zhang, L.X. Gao, G.D. Zhou, Self-assembled urea-amine compound as vapor phase corrosion inhibitor for mild steel, Surf. Coat. Technol. 204 (2010) 1646-1650. https://doi.org/10.1016/j.surfcoat.2009.10.054

[45] E. Cano, D. Bastidas, J. Simancas, J. Bastidas, Dicyclohexylamine nitrite as volatile corrosion inhibitor for steel in polluted environments, Corros. 61 (2005) 473–479. https://doi.org/10.5006/1.3280647

[46] A. Subramania, A.S. Priya, T. Vasudevan, Diethylamine phosphate as VPI for steel components, Mater. Chem. Phys. 100 (2006) 193–197. https://doi.org/10.1016/j.matchemphys.2005.12.030

[47] V. Saini, H. Kumar, Study of amine as vapor phase corrosion inhibitors for mild steel under different aggressive atmospheric conditions at high temperature, Int. J. Eng. Inn. Tech. 3 (2014) 248–256.

[48] M.A. Quraishi, D. Jamal, Synthesis and evaluation of some organic vapor phase corrosion inhibitors, Ind. J. Chem. Tech. 11 (2004) 459–464.

[49] P. Premkumar, K. Kannan, M. Natesan, Evaluation of menthol as vapor phase corrosion inhibitor for mild steel in NaCl environment, Ara. J. Sci. Eng. 34 (2009) 71-79.

[50] V. Saini, H. Kumar, DAA, 1-BIZ and 5-ATZ as vapor phase corrosion inhibitors for mild steel under different aggressive atmospheric conditions at high temperature, Int. Lett. Chem., Phys. Astron. 36 (2014) 174–192. https://doi.org/10.18052/www.scipress.com/ILCPA.36.174

Keyword Index

About the Editors

Dr. Inamuddin is working as Assistant Professor at the Department of Applied Chemistry, Aligarh Muslim University, Aligarh, India. He obtained Master of Science degree in Organic Chemistry from Chaudhary Charan Singh (CCS) University, Meerut, India, in 2002. He received his Master of Philosophy and Doctor of Philosophy degrees in Applied Chemistry from Aligarh Muslim University (AMU), India, in 2004 and 2007, respectively. He has extensive research experience in multidisciplinary fields of Analytical Chemistry, Materials Chemistry, and Electrochemistry and, more specifically, Renewable Energy and Environment. He has worked on different research projects as project fellow and senior research fellow funded by University Grants Commission (UGC), Government of India, and Council of Scientific and Industrial Research (CSIR), Government of India. He has received Fast Track Young Scientist Award from the Department of Science and Technology, India, to work in the area of bending actuators and artificial muscles. He has completed four major research projects sanctioned by University Grant Commission, Department of Science and Technology, Council of Scientific and Industrial Research, and Council of Science and Technology, India. He has published 175 research articles in international journals of repute and nineteen book chapters in knowledge-based book editions published by renowned international publishers. He has published 110 edited books with Springer (U.K.), Elsevier, Nova Science Publishers, Inc. (U.S.A.), CRC Press Taylor & Francis Asia Pacific, Trans Tech Publications Ltd. (Switzerland), IntechOpen Limited (U.K.), Wiley-Scrivener, (U.S.A.) and Materials Research Forum LLC (U.S.A). He is a member of various journals' editorial boards. He is also serving as Associate Editor for journals (Environmental Chemistry Letter, Applied Water Science and Euro-Mediterranean Journal for Environmental Integration, Springer-Nature), Frontiers Section Editor (Current Analytical Chemistry, Bentham Science Publishers), Editorial Board Member (Scientific Reports-Nature), Editor (Eurasian Journal of Analytical Chemistry), and Review Editor (Frontiers in Chemistry, Frontiers, U.K.) He is also guest-editing various special thematic special issues to the journals of Elsevier, Bentham Science Publishers, and John Wiley & Sons, Inc. He has attended as well as chaired sessions in various international and national conferences. He has worked as a Postdoctoral Fellow, leading a research team at the Creative Research Initiative Center for Bio-Artificial Muscle, Hanyang University, South Korea, in the field of renewable energy, especially biofuel cells. He has also worked as a Postdoctoral Fellow at the Center of Research Excellence in Renewable Energy, King Fahd University of Petroleum and Minerals, Saudi Arabia, in the field of polymer electrolyte membrane fuel cells and computational fluid dynamics of polymer electrolyte membrane fuel cells. He is a life member of the Journal of the Indian

Chemical Society. His research interest includes ion exchange materials, a sensor for heavy metal ions, biofuel cells, supercapacitors and bending actuators.

Dr. Mohd Imran Ahamed received his Ph.D degree on the topic "Synthesis and characterization of inorganic-organic composite heavy metals selective cation-exchangers and their analytical applications", from Aligarh Muslim University, Aligarh, India in 2019. He has published several research and review articles in the journals of international recognition. Springer (U.K.), Elsevier, CRC Press Taylor & Francis Asia Pacific and Materials Research Forum LLC (U.S.A). He has completed his B.Sc. (Hons) Chemistry from Aligarh Muslim University, Aligarh, India, and M.Sc. (Organic Chemistry) from Dr. Bhimrao Ambedkar University, Agra, India. He has co-edited more than 20 books with Springer (U.K.), Elsevier, CRC Press Taylor & Francis Asia Pacific and Materials Research Forum LLC (U.S.A) and Wiley-Scrivener, (U.S.A.). His research work includes ion-exchange chromatography, wastewater treatment, and analysis, bending actuator and electrospinning.

Dr. Rajender Boddula is currently working with Chinese Academy of Sciences-President's International Fellowship Initiative (CAS-PIFI) at National Center for Nanoscience and Technology (NCNST, Beijing). He obtained Master of Science in Organic Chemistry from Kakatiya University, Warangal, India, in 2008. He received his Doctor of Philosophy in Chemistry with the highest honours in 2014 for the work entitled "Synthesis and Characterization of Polyanilines for Supercapacitor and Catalytic Applications" at the CSIR-Indian Institute of Chemical Technology (CSIR-IICT) and Kakatiya University (India). Before joining National Center for Nanoscience and Technology (NCNST) as CAS-PIFI research fellow, China, worked as senior research associate and Postdoc at National Tsing-Hua University (NTHU, Taiwan) respectively in the fields of bio-fuel and CO_2 reduction applications. His academic honors include University Grants Commission National Fellowship and many merit scholarships, study-abroad fellowships from Australian Endeavour Research Fellowship, and CAS-PIFI. He has published many scientific articles in international peer-reviewed journals and has authored around twenty book chapters, and he is also serving as an editorial board member and a referee for reputed international peer-reviewed journals. He has published edited books with Springer (UK), Elsevier, Materials Research Forum LLC (USA), Wiley-Scrivener, (U.S.A.) and CRC Press Taylor & Francis group. His specialized areas of research are energy conversion and storage, which include sustainable nanomaterials, graphene, polymer composites, heterogeneous catalysis for organic transformations, environmental remediation technologies, photoelectrochemical water-splitting devices, biofuel cells, batteries and supercapacitors.

Dr. Mohammad Luqman has 12+ years of post-PhD experience in Teaching, Research, and Administration. Currently, he is serving as an Assistant Professor of Chemical Engineering in Taibah University, Saudi Arabia. Before joining here, he served as an Assistant Professor in College of Applied Science at A'Sharqiyah University, Oman, and in College of Engineering at King Saud University, Saudi Arabia. He served as a Research Engineer in SAMSUNG Cheil Industries, South Korea. Moreover, he served as a post-doctoral fellow at Artificial Muscle Research Center, Konkuk University, South Korea, in the field of Ionic Polymer Metal Composites for the development of Artificial Muscles, Robotic Actuators and Dynamic Sensors. He earned his PhD degree in the field of Ionomers (Ion-containing Polymers), from Chosun University, South Korea. He successfully served as an Editor to three books, published by world renowned publishers. He published numerous high-quality papers, and book chapters. He is serving as an Editor and editorial/review board members to many International SCI and Non-SCI journals. He has attracted a few important research grants from industry and academia. His research interests include but not limited to Development of Ionomer/Polyelectrolyte/non-ionic Polymer Nanocomposites/Blends for Smart and Industrial/Engineering Applications.

www.ingramcontent.com/pod-product-compliance
Lightning Source LLC
Chambersburg PA
CBHW071159210326
41597CB00016B/1606